Forest Carbon Practices and Low Carbon Development in China

Zhi Lu · Xiaoquan Zhang ·
Jian Ma · Caifu Tang
Editors

Forest Carbon Practices and Low Carbon Development in China

北京大学出版社
PEKING UNIVERSITY PRESS

Springer

Editors
Zhi Lu
School of Life Sciences
Peking University
Beijing, China

Jian Ma
Paradise International Foundation
Beijing, China

Xiaoquan Zhang
The Nature Conservancy
Beijing, China

Caifu Tang
Shanshui Conservation Center
Beijing, China

The translator of this book is Yue Meng.

ISBN 978-981-13-7363-3 ISBN 978-981-13-7364-0 (eBook)
https://doi.org/10.1007/978-981-13-7364-0

Jointly published with Peking University Press, Beijing, China
The print edition is not for sale in China Mainland. Customers from China Mainland please order the print book from: Peking University Press.
ISBN of the Peking University Press edition: 978-7-301-25125-6

Library of Congress Control Number: 2019936287

This Springer imprint is published by the registered company Springer Nature Singapore Pte Ltd.
The registered company address is: 152 Beach Road, #21-01/04 Gateway East, Singapore 189721, Singapore

Preface

Forests of Multi-benefits: Dual Implication in the Context of Low Carbon Development

The Copenhagen Climate Summit held towards the end of 2009 aroused heated discussion around the world regarding global climate change. Although the Summit didn't reach a legally binding intergovernmental agreement, it attracted people's attention to climate change and concluded more consensus of action. Global climate change is an undeniable fact, the main cause of which is attributed to greenhouse gas (GHG) emitted from the consumption of fossil fuel during industrial activities. According to Assessment Report 5 (AR5, released in 2013) of the Intergovernmental Panel on Climate Change (IPCC), global climate change is even worse than we thought in the past. Between 1880 and 2012, the average temperature of global land and sea surface is on a trend of linear increase and is up by 0.85 °C; between 2003 and 2012, annual average temperature is up by 0.78 °C based on that of 1850–1900; between 1901 and 2010, global sea level went up about 0.19 m, during which the average rising rate reached 2.0 mm per year between 1971 and 2010, and 3.2 mm per year between 1993 and 2010. Most regions witnessed shrinking snow cover, particularly in spring and summer. In the past 40 years, the amount of snow in the Northern Hemisphere diminishes monthly (excluding November and December), which is prominent during 1980s.

The consequence of global warming can't be estimated precisely for the time being, and its complexity and uncertainty may well exceed our existing knowledge boundary. Globally speaking, climate change is taking effect, such as increased desertification, change of raindrop pattern, rise of sea level and frequent drastic weather, which will deal major impact on the environment and ecosystem that human beings and animals rely upon for living. According to AR4 of IPCC, if future global temperature rise goes up by 2–3 °C compared to that of 1750 (before the Industrial Revolution), significant changes will happen to 25–40% of the existing structure and function of ecosystem on the earth. According to AR5 of IPCC, the temperature rise of 4 °C can't be avoided if no effective action is taken.

In the past century, China's climate change was in sync with global trend. China is among the countries that suffer from the impact of climate change, particularly in the northern and western part of China. Generally speaking, there are already some negative changes related to climate in China, for example, worsened and more frequent drastic weather incidents, increase in instability of agricultural production; prominent shrinking of icebergs; affected biodiversity; daunting water resources. In the recent 50 years, the amount of rainfall remains generally unchanged, but its spatial pattern has changed significantly.

Low carbon transformation is also the needed option for China's sustainable economic growth. While improving energy efficiency and promoting renewable energy, protecting biodiversity and ecology is also a key measure for combating climate change. Biodiversity is the fundamental component to a sound ecosystem of the earth in terms of species, ecosystem and genes. Ecosystem possesses dual implication of mitigation and adaptation with respect to climate change. The services provided by ecosystem, including direct product provision of food, oxygen, drug and fibre, and regulating function such as climate, flood, disease and water resource, as well as our cultural and spiritual enjoyment on the basis of the nature, are the basic security guarantee for our living. A complete, sound and diversified ecosystem can better sustain and prosper in the face of climate change. Therefore, protecting a sound ecosystem is one of the fundamental approaches to mitigating the negative impact of climate change and adapting to it. In addition, forests, grasslands and wetlands all have the capacities to absorb carbon dioxide (CO_2) in the ecosystem; as a result, there is expanding forest coverage for the benefit of CO_2 absorption on the earth is an effective measure for carbon reduction and carbon rebalance. In fact, 20–30% of global carbon emission is caused by deforestation and loss of vegetation coverage. Thus, protecting biodiversity and conserving the nature and ecosystem are even more crucial in the context of climate change.

It has been noted that China's biodiversity is faced with severe risk, which is even worse given the impact of climate change. Same as the global situation, China's biodiversity trend is deteriorating. In the recent 100–200 years, more than 10 mammals had already gone extinct in China, plus that more than 20 animals are on the verge of extinction. In particular, in the past half a century, under the aggregated effect of population growth, over-exploitation and over-utilization of resource and habitat, environmental pollution and intrusion of external species, ecosystem is becoming worse and worse. Protecting and recovering habitat may help to slow down the loss of biodiversity and reverse the trend.

With the recovery of habitat, vegetation cover may, while growing, absorb GHG and store it within the organism in the soil to create a certain amount of carbon absorption. This is what is known as carbon sink. Forest carbon may absorb GHG while promoting biodiversity protection.

In fact, forests have always been absorbing GHG and generating carbon sink. The only thing is that such fundamental eco-service provided by the ecosystem is being ignored all the time. It comes into view because it can be traded. The emergence of market trade in carbon shows that the eco-service value from carbon absorption of forests is recognized by market. Back in 2005, the public couldn't believe the notion that 'the air can be traded', but now forest carbon becomes one of the three main contents of all provinces in the policies of coping with climate change. Forest carbon is also demystified.

However, the forest carbon market is different from carbon trading in the energy and industrial sector; in that, it is quite small and the demand is very limited. In particular, with regard to Clean Development Mechanism (CDM), due to protracted negotiation in climate change and inborn deficiency in non-permanence and leakage, CDM market has low demand on forest carbon, while project development standard is strict and cost is high. Only a few forest carbon projects from China have been successfully registered in CDM. So far, only five projects have registered successfully in this regard. Meanwhile, there are not much forest carbon projects in voluntary markets. However, for the purpose of public interest, there is a rapid growth of corporate and individual funding for carbon sink projects of public interest, so as to support the restoration of forest vegetation cover for achieving carbon neutral. There is specialized public offering fund engaged in the operation.

The reason why forest carbon is favoured greatly by the public is because it can bring multiple benefits. Investing in forest carbon can simultaneously expand the habitat of wild fauna and flora and protect biodiversity, and the communities in remote areas can also benefit from forests. Forest carbon brings more value than singular carbon. Therefore, in the future, forest carbon will also enjoy greater prospect in China's carbon market. The key to China's future carbon sink market is to draw upon the experiences and lessons learned from the development of international forest carbon market, promote multi-benefit forest carbon standard and formulate carbon sink standard that is in line with China's actual condition.

This book invites experienced experts and frontline practitioners in China's carbon trading and forest carbon sector to make analysis and prediction of China's carbon market and the future of forest carbon market. It also reviews and analyses CDM projects, international voluntary market projects, domestic voluntary market projects and carbon sink projects of public interest. These real cases of carbon sink projects will help the public and decision makers to make proper judgement of the development space of forest carbon projects, so as to make better use of forest carbon to achieve a balance between conserving the ecology and combating climate change.

By the end of 2012, the first commitment period of the Kyoto Protocol finished. Global carbon market underwent significant changes. Meanwhile, China's carbon market will also start from pilots and gradually become a reality. We hope that the development experience and lessons learned from China's forest carbon projects in the past 7 years will provide guidance for the development of China's future carbon market and forest carbon. The book pools together cases of forest carbon projects that are most relevant in China, which we also hope could guide the direction of forest carbon projects. In 2015, after a difficult negotiation, the Paris Agreement was reached with milestone relevance. Low carbon and green development become a common choice of the international community; in particular, China becomes the most important player in green development and global effort in combating climate change. The output of China's experience will be even more important. For this purpose, with the joint efforts of Peking University Press and Springer and Nature, we update several chapters of this book in 2017 and publish a translated version in English.

From drafting to release, the Beijing Shanshui Conservation Center and the Nature Conservancy of US offer great support in the organization, compiling and review process. The authors work diligently on several revisions. Fangyi Yang and Shi Xiangying from the Beijing Shanshui Conservation Center have done great coordination, and Prof. Huang Wei from Peking University Press contributes a lot in terms of compilation, to whom we wish to express our sincere appreciation. Our special thanks also to Prof. Huang Wei from PUP and Prof. Huang Mengchu from Springer and Nature for their time and efforts devoted to the update and translation of the English version of the book.

Beijing, China Zhi Lu
Beijing, China Xiaoquan Zhang
Beijing, China Jian Ma
Chengdu, China Caifu Tang
January 2019

To the original version of this book, the following translator information has been added in the copyright page: "The translator of this book is Yue Meng".

Contents

China's Low-Carbon Transformation and Emergence of Domestic Carbon Market 1
Guoqiang Qian, Xiaochen Huang, Han Lai and Xiang Zou

Response to Climate Change by China's Forestry and Vision of Forest Carbon Market 47
Chunfeng Wang

General Concepts and Development Process of Forest Carbon Projects 71
Caifu Tang and Jian Ma

Reforestation Projects at Pearl River Basin of Guangxi in China 91
Sanzhong He and Zhuping Mo

Afforestation and Reforestation Project on the Degraded Land in Northwest Sichuan, China 129
Caifu Tang, Jian Ma and Biao Yang

Restoration Project of Small-Scale Reforestation in Tengchong of Yunnan Province 175
Jian Ma, Caifu Tang and Biao Yang

Public Voluntary Forest Carbon Project in China 209
Nuyun Li and Fangyi Yang

Potential of Forest Management Carbon in China 243
Wen Zhang and Caifu Tang

Obstacles, Experiences, and Recommendations on Forest Carbon Projects 269
Chunfeng Wang and Caifu Tang

Appendix 297

Editors and Contributors

About the Editors

Zhi Lu is Professor of Conservation Biology at Peking University, Executive Director of the Center for Nature and Society and Director of the Shanshui Nature Conservation Center. Since 1985, she has been engaged in the study, practice, capacity building and policy promotion in China's nature conservation.

Dr. Xiaoquan Zhang has been involved in research on the impacts of climate change on forestry and adaptation strategies, carbon accounting relevant to land use change and forestry (LUCF) activities, and methodological and project development under the Clean Development Mechanism (CDM), China Certified Emission Reduction (CCER) and Verified Carbon Standard (VCS). He has published over 100 academic papers in national and international journals and 11 books.

Jian Ma is the founder and vice president of the Paradise International Foundation. He has worked for the Nature Conservancy's Forest Carbon Program for 12 years. Under his leadership, the team has developed 4 forest carbon projects, which have restored 10,000 forests in China's biodiversity hotspots. Mr. Jian Ma also leads TNC's science team, working with the Ministry of Environmental Protection to define China's biodiversity roadmap for 2010–2030.

Caifu Tang Senior Engineer in forestry, Project Director for Southwest mountain areas of the Beijing Shanshui Conservation Center, is responsible for developing and implementing forestry carbon projects and community forestry.

Contributors

Sanzhong He Senior Engineer of Forestry Project Office of Foreign Investment of Guangxi, is responsible for developing foreign investment in forestry project, project preparation, implementation and technical training.

Xiaochen Huang Investment manager of SinoCarbon Innovation and Investment Co. Ltd., is responsible for green finance consulting, carbon finance training, investment mergers and acquisitions in the fields of ecology, energy and environmental protection.

Han Lai Analyst of SinoCarbon Innovation and Investment Co. Ltd., focus on policy research in carbon emissions trading, and management consulting in green finance and low carbon development.

Nuyun Li is Senior Engineer, Executive Deputy Head of Climate Change Office of SFGA and Secretary General of CGCF.

Zhuping Mo Senior Engineer of Guangxi Forestry Inventory and Planning Institute, is responsible for forestry planning, design and consultation, and forestry carbon measurement and monitoring.

Guoqiang Qian Strategic Director of SinoCarbon Innovation and Investment Co., Ltd., former delegate in climate change negotiation of China and expert of Asia Development Bank, is responsible for policy study and consultation in low carbon and carbon trading market.

Chunfeng Wang Senior Engineer, Deputy Head of Asian-Pacific Forestry Network Management Center of SFGA, is responsible for forestry adaptation to climate change and is SFGA facilitator of negotiation on forestry in climate change.

Biao Yang Field Project Director of Conservation International, doctoral candidate in ecology, is responsible for organizing and implementing filed work in species, fresh water, climate change, etc. He is engaged in ecology study and protective biology.

Fangyi Yang Program director at the Paradise Foundation, is responsible for conservation finance projects. He has working with Shanshui Conservation Center for Forest Carbon Projects.

Wen Zhang Senior Engineer of Sichuan Forestry Inventory and Planning Institute, is responsible for forestry resource monitoring, application of '3S' in forestry and forestry carbon project development, measurement and monitoring.

Xiang Zou Analyst of SinoCarbon Innovation and Investment Co. Ltd., focus on policy research in low carbon development and carbon emissions trading, and carbon management system development at enterprise level.

China's Low-Carbon Transformation and Emergence of Domestic Carbon Market

Guoqiang Qian, Xiaochen Huang, Han Lai and Xiang Zou

Executive Summary

Climate change is a hot topic globally. United Nations (UN) and member states reached a cooperative framework for combating climate change by reaching such legal agreements the United Nations Framework Convention on Climate Change (UNFCCC), the Kyoto Protocol, and the Paris Agreement. The Paris Agreement identifies the objective and blueprint for global low-carbon development and heralds the arrival of the global low-carbon era. Implementing the Paris Agreement and practically promoting low-carbon green development have become a major task for all countries. Currently, some major countries have formulated a series of objectives and action plans to cope with climate change. Fighting against climate change and boosting low-carbon green development are regarded key measures for increasing job opportunities and creating economic growth opportunities.

Carbon trading mechanism is a policy tool created on the basis of carbon pricing theory. This mechanism has become an important approach for many countries and regions in implementing greenhouse gas (GHG) emission reduction and combating climate change. Global carbon trading market features multilayered and diversified structure and content. Voluntary emission reduction market stands out in the global carbon trading arena and shifts to a rapid growth period, thanks to its distinctive advantage in promoting the public interest. Since the Kyoto Protocol takes effect, global carbon trading has made a great stride. The global carbon market is affected

G. Qian (✉) · X. Huang · H. Lai · X. Zou
Sino Carbon Ltd., Qingdao, China
e-mail: qianguoqiang@sino-carbon.cn

X. Huang
e-mail: huangxiaochen@sino-carbon.cn

H. Lai
e-mail: laihan@sino-carbon.cn

X. Zou
e-mail: zouxiang@sino-carbon.cn

Z. Lu et al. (eds.), *Forest Carbon Practices and Low Carbon Development in China*, https://doi.org/10.1007/978-981-13-7364-0_1

by policy uncertainty around the world and weak market demand, and the Kyoto Protocol-based global carbon market is lost in the quagmire. Entering 2012, the market encountered decreasing trading price and trade volume of products. However, with the Paris Agreement comes into view, the global carbon market is coming back, with more and more countries and regions promoting the establishment of a domestic carbon trading market. The future of carbon trading is bright.

Achieving low-carbon and green development and low-carbon transformation is an integral part of China's ecological civilization. It is the necessary choice for China's sustainable economic and social development. China formulates national plan and work plan to cope with climate change, conducts low-carbon pilot and incorporate "green development" in the economic and social development plan, establishes founding principle, specific targets, key areas, policy measures and steps in combating climate change, and gradually creates and improves a management regime and work mechanism for climate change.

Currently, local carbon market is operating as had been expected. It uses the market mechanism to achieve China's action objective of controlling GHG emission at a lower cost, and lays a solid foundation for the gradual establishment of a unified carbon market. By drawing upon the experience and lessons learned from 7 carbon emission trading pilots, the national carbon market started in 2017, with related development work already underway in an orderly fashion.

Forest carbon plays an irreplaceable role in mitigating global climate change and improving ecology and environment. It enjoys a broader prospect as long as it receives policy support. Currently, China's forestry project serves two major purposes, one is being used as an offset product in mandatory carbon market; the other is being used as carbon-neutral product in voluntary emission reduction trade.

1 Political Process of Combating Climate Change Globally and Low-Carbon Development

1.1 Process of World Climate Politics

(I) **Global climate change and attribution**

In order to scientifically assess the cause, impact, and countermeasures for climate change, as mandated by the United Nations (UN), the World Meteorological Organization (WMO) and the United Nations Environment Programme (UNEP) jointly set up the Intergovernmental Panel on Climate Change (IPCC) in 1988. The purpose is to, on the basis of comprehensiveness, objectivity, openness and transparency, assess and understand the relation between anthropogenic activity and climate change, the potential impact of climate change and scientific and technological and social and economic information that are relevant to the scientific basis of adapting to and mitigating climate change.

Table 1 Main assessment conclusion of IPCC with respect to climate change

1. During the period 1880–2012, the average surface temperature increased by 0.65–1.06 °C, showing a prominent trend of global warming
2. Anthropogenic activity is the main cause of the increased concentration of GHG in the atmosphere. The credibility of this conclusion is over 95%
3. Climate change subjects ocean system, the majority of land and freshwater species to higher risk of extinction, coastal and low-lying areas suffer higher ecological pressure, and food security is reduced. These severely threaten the health and security of human beings
4. Given the current trend, if no further action is taken, the concentration of CO_2 in the atmosphere will exceed 750 ppm by end of the century, the average surface temperature will be 3.7–4.8 °C higher than that in the Industrialization period. This cause greater and more frequent extreme weather incidents and disaster
5. In order to avoid the detrimental impact of climate change, temperature increase shall be controlled within 2 °C by the end of the century. By 2050, global carbon emission needs to be reduced by 40–70% on the basis of that in 2010

As the most authoritative assessment agency in climate change, the IPCC, since its establishment in 1988, organizes thousands of top experts around the world to have compiled five assessment reports on climate change. The latest assessment report was finished in 2014. Its main conclusion is shown in Table 1.

With each assessment report (AR), IPCC makes a crucial contribution to the negotiation in climate change and international cooperation. For example, the AR 1 (released by IPCC in 1990) directly promotes the UN to create a negotiation committee on climate change convention, which leads up to the UNFCCC. Shortly before the COP21, IPCC released the AR 5. The up-to-date scientific evidence and assessment conclusion in this report offers strong support for reaching the Paris Agreement (Table 2).

(II) **History of the global climate regime**

In order to effectively deal with the threat posed by climate change, the international community began to seek countermeasures since the 1990s. In December 1990, the UN General Assembly (UNGA) adopted Resolution 45/212, deciding to establish an intergovernmental committee on climate change under the leadership of UNGA to negotiate and draft a framework convention on climate change. On May 9, 1992, the committee adopted the UNFCCC in New York. As the first legal instrument on climate change, the Convention identifies the ultimate goal of stabilizing GHG concentration in the atmosphere on a level that can avoid climate change be disturbed by risky anthropogenic activities. It also sets out that developed countries and developing countries shall undertake "common but differentiated responsibilities" in coping with climate change.

Table 2 IPCC assessment report and progression of climate negotiation

	IPCC AR 1 (1990) The UN set up intergovernmental negotiation committee, and drafted UNFCCC	UNFCCC came into effect (1994) Establish a global objective of combating climate change, identify the responsibilities of main participants, lay down legal foundation the basic framework of international cooperation on climate change
	IPCC AR 2 (1995) Creation of the Kyoto Protocol	The Kyoto Protocol is approved (1997) Set quantified emission obligation for developed countries, namely, reducing GHG emission by 5% on the basis of 1990 during the first commitment period (2008–2012)
	IPCC AR 3 (2001) Creation of the Marrakesh Agreement	The Kyoto Protocol came into effect (2005) The Marrakesh Agreement creates a complete set of approach to monitoring and calculating the emission of developed countries in implementing the Kyoto Protocol, as well as measures for accomplishing the objectives and evaluation mechanism
	IPCC AR 4 (2007) Initiation of Bali Roadmap negotiation	Bali Roadmap is launched (2007) Discussion of emission reduction mechanism after the first commitment period of the Kyoto Protocol expires in 2012 Copenhagen Accord (2009) Pooling of key political consensus for a cooperation mechanism between 2012–2020 Durban Platform is launched (2011) Initiation of emission reduction negotiation for post 2020
	IPCC AR 5 (2014) Creation of the Paris Agreement	Paris Agreement (2015) Global emission reduction agreement for post 2020, key milestone of combating climate change globally, the strong signal of low-carbon transformation of global economy Marrakesh Action proclamation (2016) Reiterate that the Paris Agreement will be fully implemented, in light of the irreversible trend of climate change. A consensus is reached on a procedural matter concerning the Agreement

The Kyoto Protocol in 1997 is the first emission reduction agreement of the international community that lays down a quantified target for specific developed countries. On the basis of the Convention, the Kyoto Protocol further demands developed countries, as listed in the annex to the Convention, to reduce the GHG emission by 5% on the basis of 1990 during the first commitment period of 2008–2012 and reduce it by 18% on the same basis during the second commitment period of 2013–2020. Specific quantified GHG emission indicators have been identified for each country in the Annex I (see Table 3).

Table 3 Quantified national GHG emission indicators of Annex I countries

Parties	Quantified limit or emission reduction indicators (2008–2012) (base year or baseline percentage)	Quantified limit or emission reduction indicators (2013–2020) (base year or baseline percentage)
Australia	108	99.5
Austria	92	80
Belarus		88
Belgium	92	80
Bulgaria	92	80
Croatia	95	80
Cyprus		80
Czech Republic	92	80
Denmark	92	80
Estonia	92	80
The European Union	92	80
Finland	92	80
France	92	80
Germany	92	80
Greece	92	80
Hungary	94	80
Iceland	110	80
Ireland	92	80
Italy	92	80
Kazakhstan		95
Latvia	92	80
Liechtenstein	92	84
Lithuania	92	80
Luxemburg	92	80
Malta		80
Monaco	92	78

(continued)

Table 3 (continued)

Parties	Quantified limit or emission reduction indicators (2008–2012) (base year or baseline percentage)	Quantified limit or emission reduction indicators (2013–2020) (base year or baseline percentage)
The Netherlands	92	80
Norway	101	84
Poland	94	80
Portugal	92	80
Romania	92	80
Slovakia	92	80
Slovenia	92	80
Spain	92	80
Sweden	92	80
Switzerland	92	84.2
Ukraine	100	76
UK	92	80
Canada	94	
Japan	94	
New Zealand	100	
Russia	100	

The Kyoto Protocol is an active attempt of global emission reduction mechanism. Since it has a bearing on economic transition and clean energy alternative, plus that it touches upon energy security, competitiveness protection, and profound economic and social factors, countries are prudent in controlling GHG emission. As the first commitment approaches, the international community initiated negotiation on "post Kyoto Protocol" emission reduction. Since there are numerous conflicts among several vested parties within developed countries, between developed and developing countries and even within developing countries, the negotiation is met with difficulties. The negotiation on the second commitment period of the Kyoto Protocol started in 2005 and protracted until the Paris Agreement in 2015, lasting for a decade.[1] Climate change becomes a global hot topic in this process.

In order to pool political consensus, the UN Secretary-General hosted several leadership summits on climate change. More than 160 heads of state and government leaders attended the COP15 Copenhagen in 2009 and the UN Climate Change Conference Paris 2015 (COP21). After a decade of heated negotiation, major countries begin to realize that promoting technology innovation and embracing energy

[1]With respect to the Paris Agreement, including the historical background, process, main achievement and comment of climate change negotiation, please refer to From Copenhagen to Paris—Change and Development of Global Climate Regime (Tsinghua University Press), co-authored by Zhu Songli and Gao Xiang.

transition and green low-carbon growth is the trend. Any delay in this regard will not only endanger our future, but also lose opportunities and advantages in new economic transition. This major change of mind finally contributes to the political foundation of the Paris Agreement.

The Paris Agreement at the end of 2015 takes less than 1 year from adoption to entry into force, while the Kyoto Protocol did the same in 8 years. The Paris Agreement is the fastest treaty that enters into force in the history of the UN, highlighting that the international community has the political determination and effort to speed up low-carbon transformation. The Paris Agreement creates a global emission reduction mechanism that all countries can partake, making itself a new milestone in combating climate change. On one hand, the Paris Agreement identifies the direction and sends a strong political signal to global low-carbon transformation, in particular, it sets the long-term goal of "achieving carbon neutral by the second half of next century", meaning that we will put an end to fossil fuel energy and move into an era of low-carbon and clean energy. On the other, the Paris Agreement sets the principle that all countries shall act without backtrack. On the basis of this principle and the evaluation mechanism of reviewing global emission reduction every 5 years, there is mechanism assurance in place for all countries to gradually improve on their action plans, so that global emission reduction efforts will be continuously going upwards.

(III) **Future trend of global climate regime**

The Paris Agreement will completely replace the Kyoto Protocol after 2020. Its mechanism will continue and connect to the norms and mechanism under the Kyoto Protocol, with its reporting, transparent rules, and market mechanism worth special attention. After the Paris Agreement came into effect, there is still a lot of work to be done, this is because first, parties are yet to put everything in detail due to time constraint; second, there exist vague expressions in the text; in addition, parties will have a series of negotiations on its implementation and funding issues.

In essence, climate change is related to the economic competitiveness of all countries and their conflicts of interest over development space and global image. In the context of a new round of change and adjustment to the international landscape, developed countries also try to further maintain and consolidate their leading role in global regime by leading global climate change negotiation and identifying a positive global climate regime for the future. However, as the US backs out of the Paris Agreement, the EU falls short of long-term strength, and global climate order and landscape are ever changing, global climate governance is entering into a "3.0 era" that is completely different from the past. China will play an increasingly important role in the reform of global social governance mode.

Whatever the implementation outcome of the Paris Agreement, it can be expected that the international community will continue with the political process of international cooperation on combating climate change, and it will be intensified continuously. Meanwhile, as the outcome of climate change negotiation, low-carbon development concept is also widely accepted by all, and has become the main guide for countries to engage in domestic actions on coping with climate change.

1.2 *Low-Carbon Transformation Becomes a Key Strategy of Development for All Countries*

Given the impact and boost of climate change negotiation, main countries formulate a series of actions and objectives targeting climate change. Countries regard climate change and low-carbon and green development as key measures for increasing new job opportunities and creating new economic growth point, in an attempt to promote sustainable economic growth and enhancing long-term competitiveness. Table 4 lists actions of major developed and developing countries in coping with climate change and promoting low-carbon development.

Table 4 List of emission reduction targets and measures

Countries	Emission reduction targets	Measures
EU	Reducing 40% by 2030 in comparison with 1990	Implementing the Act on 2030 Climate Change and Energy, planning to include carbon emission trading sector and non-trading sector in the framework of 2030 climate and energy. Further reforming and adjusting the operating scheme of carbon trading market
US	Reducing 26–28% by 2025 in comparison with 2005	Implementing Clean Air Act, Energy Policy Act, and Energy Independence and Security Act, improving traffic and construction efficiency and giving full play to alternative energy. All states will also take actions, especially California will take the lead in initiating carbon trading mechanism
Japan	Reducing 26% by 2030 in comparison with 2013 (25.4% in comparison with 2005)	Implementing Act on Promoting Countermeasures against Global Warming and Basic Act on Energy Policy, and formulating laws on energy conservation and emission reduction in all sectors. Starting market-oriented emission reduction and taking the lead in launching carbon market in Asia
Australia	Reducing 26–28% by 2030 in comparison with 2005	Setting up Emission Reduction Fund (ERF), giving full play to renewable energy, improving the efficiency of existing energy, and promoting "national climate adaptation strategy". Adopting carbon trading act in 2012. Though abolished by conservative government in 2014, it may be started over again in the foreseeable future

(continued)

Table 4 (continued)

Countries	Emission reduction targets	Measures
India	In 2030, carbon emission per unit GDP is down by 33–35% in comparison with 2005	Implementing National Environment Policy, National Action Plan on Climate Change, and Action Plan on Climate Change at State-level, setting up clean and efficient energy system, improving industrial energy efficiency. Working on the market mechanism for GHG emission reduction
Brazil	Reducing 37% by 2025 in comparison with 2005	Adopting the National Climate Change Act, Natural Forestry Protection Law, and National Land Use Act, which serve as the basis for carbon emission reduction. Implementing "National Adaptation Plan (NAP)" to cope with the potential negative impact of climate change. Working on the market mechanism for GHG emission reduction
South Africa	GHG emission will reach the peak during 2020–2025, and begin to drop after a decade, namely in 2035. GHG emission in 2020 will be down by 34% compared to BAU, and down by 42% by 2025	Discussing to introduce carbon tax scheme to set emission criteria for key companies, so that those beyond the standard will be levied with tax. Companies are allowed to purchase the quantity of carbon emission reduction from the global carbon market to offset their tax. A series of policies supporting green industries and renewable energy development have been released, including "protective price for renewable energy", "government subsidy plan for renewable energy", "market conversion project for renewable energy", "certificate trade of renewable energy", and "wind power project of South Africa"
Republic of Korea	In 2030, the emission is down by 37% compared to BAU	Establishing Ministry of Environment (MOE), and Ministry of Trade, Industry and Energy (MOTIE) as the authorities of emission reduction. Releasing "GHG emission reduction roadmap" in 2014, laying down emission reduction plan and implementation approach for all sectors, and making further reform and adjustment to the operating scheme of carbon trading market

(continued)

Table 4 (continued)

Countries	Emission reduction targets	Measures
Mexico	In 2030, the emission is down by 25% compared to BAU	Releasing the Climate Change Act in 2012, and National Strategy in Combating Climate Change in 2014, making plans for the coming 10, 20 and 40 years in climate strategy. The carbon tax will be levied on a pilot basis from 2013

(I) **Low-carbon policies in the EU and UK**

1. **The EU**. EU is the r and is leading low-carbon policy. As a key participant of global carbon market, the EU drafted adequate legislations to promote low carbon development at an earlier date.

The EU adopted a package plan for energy and climate in 2010. It sets up a series of action targets for combating climate change by means of legislation, including reducing GHG emission by 20% by 2020 on the basis of 1990, improving energy efficiency by 20% by 2020, and improving the use of renewable energy by 20% in total energy consumption.

Based on the above targets, the European Commission released the 2030 Framework for Climate and Energy Policy (hereinafter as "the Framework") on January 22, 2014, putting forward EU's target for interim GHG emission and percentage of renewable energy by 2030. The contents of the Framework cover the following:

(1) Before 2030, reduce EU's GHG emission by 40% as compared to 1990. If this is achieved, the emission of sectors involved in EU's emission trade system will have to reduce by 43% as compared to 2005, while those not involved will have to reduce by 30% as compared to 2005. This target will be broke down to be undertaken by member countries.
(2) Before 2030, the percentage of renewable energy will be 27% of total energy consumption. The EU will not adopt legislation to set obligatory regulations on its member countries. They can make flexible adjustment according to the situation and conditions of their own energy systems.
(3) Reform EU's emission trading system. The European Union Emission Trading Scheme (EU ETS) started operation on January 1, 2005. It is the most important policy tools for EU to achieve emission reduction target, and it is also the largest emission trading system in the world. In order to make it more effective in attracting low-carbon industrial investment, the European Commission suggests reforming the market stable reserve mechanism to deal with the surplus in the European Union Allowances (EUAs), so as to improve system flexibility and respond to supply regulation with respect to allowance accounting.

2. **UK**. UK is an active promoter and forerunner in the global low carbon economy. It not only creates business and job opportunities by means of developing, applying, and exporting low-carbon technology, but also plays a leading role for Europe and the world at large in terms of low-carbon economic transformation.

In 2008, UK government formulated Climate Change Act (CCA), which committed UK to reducing GHG by 80% by 2050 as compared with 1990, and identifies "carbon budget" for the coming 5 years. This is the very first act with identified GHG emission reduction target. It also makes UK the first country in the world to embrace a legally binding and long-term structure with respect to GHG emission reduction and climate change adaptation.

Shortly after the "Brexit vote", the UK government declared new CO_2 emission reduction target, which plans to reduce carbon emission by 57% by 2032 as compared to 1990 and submits the fifth "carbon budget" to the parliament. Those who are concerned that "Brexit" may affect the UK's climate change policy can now be relaxed, but the new emission reduction target implies that the government will work up to draft a new energy strategy that is consistent with cost–benefit, reducing the reliance on imported natural gas and greatly cuts energy demand of UK families.

(II) **Low-carbon policies for Umbrella Group countries**

US, Japan, Australia, and some developed countries form up the "Umbrella Group" in the climate change negotiation. Though not as active as the EU on climate change policy, they also take actions domestically against climate change.

1. **US**. US is swinging from side to side with regard to climate change, due to party politics. In 2015, the Obama government, in its submission to the UN, noted that its Intended Nationally Determined Contribution (INDC) in 2025 will be 26–28% less compared to 2005, and ratified the Paris Agreement in 2016. Since Trump took office in 2017, the US began to adopt a negative policy on climate change and provided to the UN an intention letter of withdrawing from the Paris Agreement.

In spite of that, some US states have been very active in climate change. In response to Trump's withdrawal from the Paris Agreement, California, New York and Washington joined together as the climate alliance, stating that they will continue to fulfill US's commitment of the year, that is, to cut the GHG emission of 2025 by 26–28% on the basis of 2005, thus reaching or surpassing federal standard for clean energy plan. In addition, mayors from 61 US cities made joint declarations on June 1, 2017 to maintain and enforce the Paris Agreement.

California has been taking the lead in combating climate change. As early as 2006, California passed the Global Warming Solutions Act (AB32) to counter global warming, and identified the reduction target of cutting GHG emission to the level of 1990 by 2020. To achieve this target, AB32 authorize California Air Resources Board (CARB) to draft emission reduction measures including market mechanism. Currently, California has passed local laws introducing GHG quantity control and emission trade scheme. Its carbon trading mechanism was officially initiated in 2013.

2. **Japan**. Japan made the commitment to cut GHG emission in 2020 by 25% on the basis of 1990 during the COP15 in 2009, and it put forward a long-term target that GHG emission by 2050 will be down by 60–80% on the current basis. However, following the Fukushima accident in March 2011, nuclear power policy was affected. The Japanese government was forced to readjust emission target, and withdraw from the second commitment period from the Kyoto Protocol during the COP18 in Doha in 2012. In its INDC submitted to the UN in 2015, Japan noted a new target that emission will be cut by 26% in 2030 compared to 2013 (cut by 25.4% compared with 2005). The year 2013 was selected as the base year, because the earthquake caused nuclear power plants to stop operation, thus GHG emission of that year reached a historical high.

In addition, there are three aspects of climate change policies and actions that worth mentioning:

First, Japan promulgated the Act on Promoting of Global Warming Countermeasures in 1998, which lays down the plan of emission reduction for all sectors under the Kyoto Protocol. Although Japan withdraws from the second commitment period of the Kyoto Protocol, it continues to amend the Act in 2013. The action plan under the Kyoto Protocol is revised as a plan to avert global warming and is still effective.

Second, due to domestic pressure from various interests groups, Japan indefinitely postponed the plan to set up a national carbon trading mechanism, but Tokyo and Saitama Prefecture started a regional carbon market in 2010 and 2011, respectively. The market in Saitama was connected immediately with that of Tokyo at the onset, allowing the flow of approved carbon credit. By September 2016, emission reduction credit totaling 5,600 tons of CO_2 equivalent flows from Saitama to Tokyo carbon market.

Third, bilateral emission reduction mechanism is confirmed to replace clean development mechanism under the Kyoto Protocol as a new mechanism for promotion. Japan is accelerating the establishment of bilateral emission reduction mechanism with Southeast Asian countries and developing countries, so that Japanese companies may obtain emission reduction credit by investing in clean technologies (such as renewable energy). By October 2017, Japan has signed bilateral agreements with Mongolia, Vietnam, Indonesia, Philippines, and other 13 countries.

3. **Australia**. Australia has adequate implementation framework and management system regarding climate change policy. Climate change issue is managed centrally. The former "Australia GHG Office", which manages climate change, was upgraded into Climate Change Ministry in 2009 with the responsibility of drafting domestic climate change policy.

On November 8, 2011, Australia Senate adopted a package plan including the Clean Energy Act, and it identified a national emission reduction target of cutting GHG by 5–25% on the basis of 2000 in 2020. In addition, Australia submitted its

INDC in August 2015, putting forward a target of cutting GHG by 26–28% in 2030 as compared to 2005. Besides, in December 2015, Australia adopted the National Energy Efficiency Plan (2015–2030), noting that the energy efficiency in 2030 will be up by 40% compared with 2015.

Furthermore, the Australia Labor Party government once introduced a carbon pricing mechanism and was officially implemented on July 1, 2012, then is abolished by the Conservative Party in 2015. The Australian government set up the Emission Reduction Fund (ERF) in 2016 to purchase emission reduction credit from compliant emission reduction projects. The system also includes a "baseline-offset" mechanism. Facilities emitting over 100,000 tons of CO_2e are required to control their emission within a threshold, thus any exceeding part needs to be offset by purchasing emission reduction credit.

(III) **Low-carbon policies in main developing countries**

1. **India**. India submitted its INDC in 2015, stating the target of reducing per unit GDP emission in 2030 by 33–35% on the basis of 2005. In order to achieve this target, India launched the National Action Plan on Climate Change, which identifies 8 core national plans for implementation, including solar energy plan, energy efficiency plan, sustainable living environment plan, water resource plan, Himalaya eco-conservation plan, green India plan, sustainable agricultural development plan and study plan on climate change strategy. In addition, 32 states (co-dependencies) have joined the State Action Plan on Climate Change, so that climate change initiatives will be promoted at the state level.
2. **Brazil**. In its INDC, Brazil set the target of cutting the emission reduction in 2025 by 37% compared to 2005. The main emission source in Brazil is deforestation, thus the key measures for controlling GHG in the future lies in reducing deforestation. China's National Plan for Combating Climate Change National Plan on Climate Change put forward a series of measures to cope with climate change in 2008, mainly including improving energy utilization efficiency; maintaining a high level of power supply by renewable energy; further utilizing biofuel; continuously reducing deforestation. On the basis of that, the Natural Forestry Protection was adopted in 2012 to further cut down on GHG emission caused by deforestation. In addition, the Brazilian government also began to implement the National Adaptation Plan since 2016 to deal with the negative impact brought on by climate change.
3. **South Africa**. South Africa set down its mid and long-term GHG emission reduction target as follows: cutting emission by 34% in 2020 compared to "baseline emission scenario", and cutting the emission by 42% in 2025. The emission will reach a peak between 2020 and 2025, and will begin to drop after keeping on that level for a decade.

In order to achieve emission reduction target, South Africa has released a series of policies in support of green industries and renewable energy, including "protective price for renewable energy", "financial subsidy plan for renewable energy", "market conversion project for renewable energy", "certificate trade of renewable energy", and "wind power project of South Africa". Meanwhile, South Africa commences to introduce carbon tax. Its taxation and budget report of 2012–2013 provided explanations and interpretations with regard to the carbon tax proposal. This proposal is similar to the fixed carbon tax in Australia, levied as of July 1, 2012. The proposal was postponed several times, and the new plan will be implemented from 2017.

4. **The Republic of Korea**. Considering its long-term economic competitiveness, the Republic of Korea has identified green and low-carbon growth as its development strategy. Korea set the target of cutting emission by 37% in 2030 in comparison with "baseline emission scenario", and incorporated this target in the INDC.

In order to achieve that aim, the Republic of Korea promulgated the Basic Law on Low-carbon and Green Growth in 2010, which is the first of its kind in the world regarding low-carbon green growth. The law lays down a fundamental regime structure for the Republic of Korea to shift toward low-carbon and green growth. On May 2, 2012, National Assembly of the Republic of Korea adopted an act introducing carbon trading mechanism, making itself the first Asian country to embrace carbon trading legislation. The Republic of Korea's trade mechanism was officially launched in 2015, covering 60% of the Republic of Korea's total emission. In December 2016, the Republic of Korea government released the Basic National Roadmap for GHG Emission Reduction by 2030, which summarized the key role of the carbon market in achieving the Republic of Korea's INDC, and broke down national targets into specific sectors.

5. **Mexico**. Mexico parliament adopted the General Law on Climate Change. This law has the main objective of promoting GHG emission reduction and effectively mitigating the negative impact of climate change. It also identified a long-term target that by 2050, Mexico's GHG emission will be 50% less than in 2000. Following the UK, Mexico is the second to set a legally binding emission target for the long term. In order to accomplish that, Mexico also set a mid-term target of cutting emission by 25% in 2030 compared to "baseline emission scenario".

In actual practice, Mexico began to try out with the carbon tax policy in 2013. On August 15, 2016, the Ministry of Environment and Natural Resources of Mexico, the Mexican Stock Exchange, and its trading platform of voluntary carbon emission reduction signed a cooperation agreement to jointly pilot on the voluntary carbon emission trading system. A simulation carbon trading operation was officially initiated in November 2016 to familiarize stakeholders with emission trading concept, raise the preparation awareness of companies to this policy tool. The simulation didn't introduce trade parameters and is expected to finish in December 2017. By then a national registration system for major companies under emission control will be online, which will lay the foundation for Mexico to execute nationwide carbon market in 2018.

2 Carbon Market—Booster for Low-Carbon Transition

2.1 Fundamentals of Carbon Trading

The core component of achieving emission target through economic means lies in setting a carbon emission price. If a company undertakes no social or economic responsibility for environmental damage caused by CO_2 emission, it will never consider carbon emission in its decision-making process, nor will it be encouraged to cut carbon emission. This is called Economic Externality. If carbon emission price is set, companies will begin to pay for its CO_2 emission, and they will consider carbon emission cost into overall financial accounting and management decision-making. They will also be stimulated to cut CO_2 emission to reduce operating cost. Meanwhile, with a sound carbon pricing mechanism, companies won't be afraid of cost increase due to technology upgrade or business improvement and potential loss of competitive advantage. This will inspire technology and management innovation.

As shown in Figs. 1 and 2, carbon trading mechanism is a policy tool created on the basis of carbon pricing theory. Government introduces the Cap and Trade system to set a limit for carbon emission. If a company's emission in a given period exceeds that set by the government, it needs to purchase the corresponding allowance through carbon trading in the market to achieve the set target, otherwise, it will be faced with a high penalty fee. A company may also choose to cut carbon emission by low-carbon technology upgrade or business improvement, or sell its allowance surplus to acquire economic benefit through carbon trading. Each company will choose a carbon emission standard favorable to itself according to its own condition and based on cost-benefit accounting, either by self-emission or by purchasing emission allowance through carbon market. Compared with a rigorous and inflexible administrative approach, this marketized measure provides companies with flexible and low-cost means, which also reduces the overall social cost for GHG quantity control.

2.2 History of International Carbon Market

Initiated by the US, the Kyoto Protocol was adopted during the 3rd Conference of Parties to the Convention held in Kyoto, Japan in December 1997. The Kyoto Protocol introduced three flexible mechanisms that are intended to help developed countries accomplish their emission reduction targets. This opens up a new channel

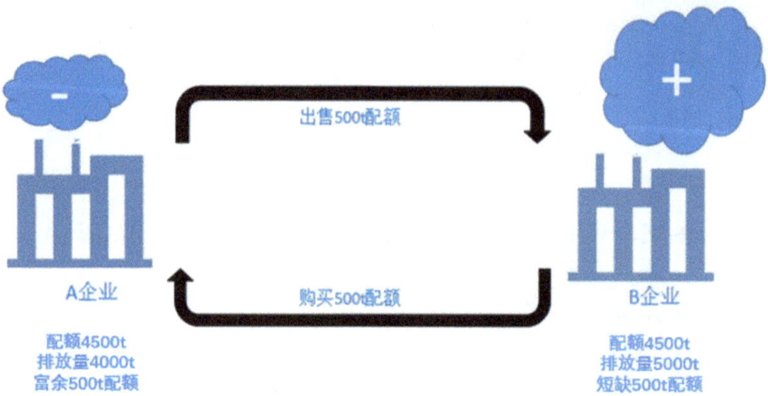

A 企业	Company A
配额 4500t	Allowance 4500t
排放量 4000t	Emission 4000t
富余 500t 配额	500t surplus
出售 500t 配额	Sell 500t allowance
购买 500t 配额	Purchase 500t allowance
B 企业	Company B
配额 4500t	AllowanceAllownance 4500t
排放量 5000t	Emission 5000t
短缺 500t 配额	500t allowanceallownance deficit

Fig. 1 Functional diagram of carbon market

for global carbon emission trading and serves as the base for global carbon trading market.

The Kyoto Protocol sets down GHG emission reduction targets for the first commitment period of 2008–2012 for the developed countries with emission reduction obligations. According to the Kyoto Protocol, by 2012, all developed countries have to cut the mission of 6 kinds of GHG.[2] (including CO_2) by 5.2% compared to 1990, while developing countries shoulder no emission reduction obligations. Emission reduction targets for developed countries are shown in Table 5.

By means of emission reduction targets, each developed country is given an assigned amount of units (AAUs), which is comparable to an assigned emission allowance by the UN. Developed countries with emission obligations can directly buy and sell or transfer their AAUs, which is the emission trading (ET) stipulated by

[2] CO_2, CH4, N_2O, HFCS, PFCS, and SF_6.

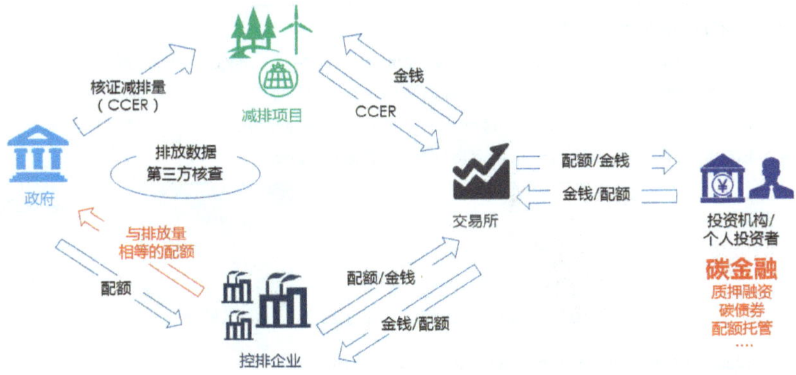

政府	Government
核证减排量（CCER）	China Certified emission reduction (CCER)
减排项目	Emission reduction projects
金钱	Money
交易所	Exchange
配额/金钱	Allowance/Money
金钱/配额	Money/Allowance
控排企业	Emission control company
与排放量相等的配额	Allowance equivalent to emission
配额	Allowance
排放数据	Emission data
第三方核查	Third party verification
投资机构/个人投资者	Investment institution/ individual investor
碳金融	Carbon finance
质押融资	Pledge financing
碳债券	Carbon bond
配额托管	Allowance Allowancetrusteeship

Fig. 2 Operating mode of carbon market

Table 5 Emission reduction targets of developing countries under the Kyoto Protocol

Countries or regions	Emission reduction targets
EU	8%
US	7%
Japan	6%
Canada	6%
East European countries	5–8%
New Zealand, Russia, and Ukraine	Stabilize on the level of 1990
Ireland, Australia, and Norway	Increase respectively by 10%, 8%, and 1%

Carbon trading created by the Kyoto Protocol

- Developed countries undertake legally-binding quantified emission reduction indicators.
- The Kyoto Protocol is founded on AAUs, ERUs and CERs.
- In principle, developed countries shall not emit beyond their AAUs.
- ERUs and CERs are also allowed for compliance purpose.

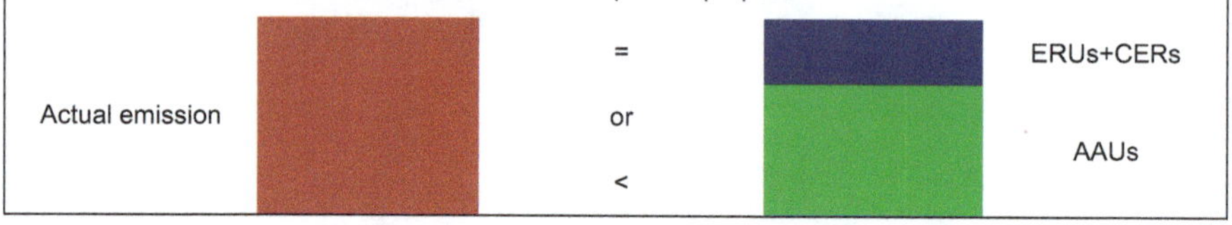

Fig. 3 Fundamentals of compliance with the Kyoto Protocol

the Kyoto Protocol. Meanwhile, the Kyoto Protocol also provides developed countries with CDM and JI as two offset mechanisms. The main difference between these two is that, CDM uses certified emission reduction (CER) as the emission reduction credit, and it comes from emission reduction projects of developing countries with no emission reduction obligations; JI use emission reductions units (ERU) as the emission reduction credit, and it comes from emission reduction projects of developed countries with emission reduction obligations.

According to the implementation of the Kyoto Protocol, developed countries with emission reduction obligations have to calculate their GHG emission quantity based on UN's accounting rules for the first commitment period, which shall not exceed their AAUs, ERUs, and CERs combined, otherwise they will be subject to penalties (Fig. 3).

In addition, EUETS also connect with the three flexible mechanisms under the Kyoto Protocol by "linking directive". Since 2005, companies under EUETS may partially use CERs and ERUs generated respectively from CDM and JI projects to offset their emission. On one hand, this expands the depth and width of the global carbon market established by the Kyoto Protocol, on the other, it also provides greater flexibility for EU ETS.

2.3 *The Classification of Carbon Market*

There exist several types of carbon market globally, with diversified levels and multiple categories in their structures and contents. Based on whether a carbon market is founded on legal basis and motive of trade, there are mainly two types of market, namely, mandatory carbon market and voluntary emission reduction market. Another way of categorization is based on the types of trading products, which covers the allowance market and offset market.

(I) **Mandatory carbon market and voluntary emission reduction market**

- Mandatory carbon market

Mandatory market is based on laws and regulations. Those participating in the market must comply with the Convention, while market demand comes from legal enforcement. The mandatory market in its early form is exhibited as CDM, JI, and ET founded by the Kyoto Protocol. Serving as a bond, the Kyoto Protocol connects main countries to create a global carbon market, which is normally known as the "Kyoto" carbon market.

Another type of mandatory market is founded on domestic legislation, including country-level carbon markets in EU, New Zealand, Republic of Korea, and local carbon markets like California, Quebec, and such provinces in China as Guangdong, Hubei, Beijing, Shanghai, Tianjin, Chongqing, and Shenzhen.

By means of legislation, EU single-handedly connects the EU carbon market with the "Kyoto" carbon market. Since 2008, companies under the EU carbon market may use CERs and ERUs to offset their mission for compliance assessment. This means that EU individually confirms CDM and JI as a legitimate offset mechanism, which further expands the demand for CERs and ERUs under the Kyoto carbon market. It is with CDM that China participated in the global carbon trading market. The quantity of emission reduction from domestic emission reduction projects is developed into CERs that are sold to companies under EU carbon market for Convention implementation, or sold to the governments of EU member states, Japan, Canada, New Zealand and other developed countries, in order for them to fulfill the first commitment period of emission reduction under the Kyoto Protocol.

- Voluntary emission reduction market

Following the principle of self-discipline, some Non-Governmental Organizations created offset mechanisms that provide CERs and certification service. These CERs usually comes from corporate social responsibilities (CSR) rather than legal enforcement. This type of carbon market is known as the voluntary emission reduction market, with its widely accepted forms of verified carbon standard (VCS) and gold standard (GS). Since the demand is not created by legal enforcement, the scale of voluntary emission reduction market is way less than mandatory markets. But voluntary markets are also less affected by policies, thus they are more stable than mandatory markets.

Voluntary emission reduction markets started earlier than mandatory markets. The earliest trade in voluntary emission reduction happened in the 1990s. The society, individual and non-profit entities, in order to voluntarily offset their GHG emission, purchased emission reduction credits generated from GHG offset projects. Around 2002, more and more companies and individuals paid attention to global

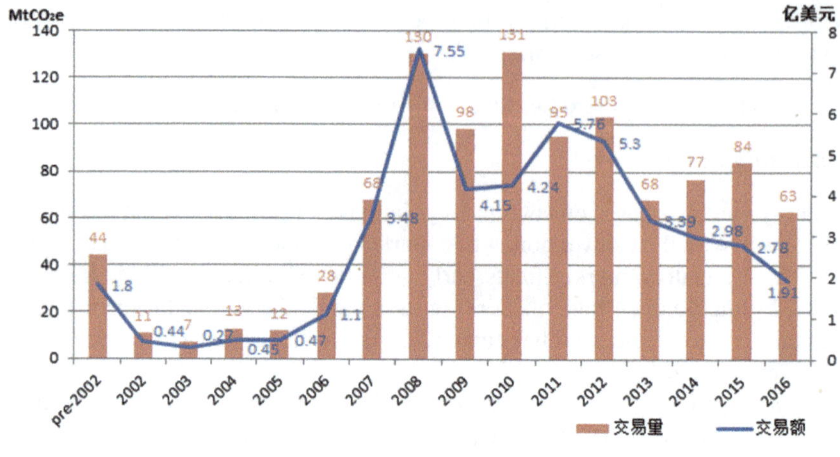

亿美元	US$100 million
交易量	Trade quantity
交易额	Trade volume

Fig. 4 Historical trade volume and quantity of global voluntary emission reduction market

warming phenomenon caused by anthropogenic activities, since then voluntary emission reduction market began to move onto the fast track. The trade volume of global voluntary emission reduction markets is shown in Fig. 4.

(II) **Allowance market and offset market**

Allowance markets trade in allowance, which is the most fundamental and principal product for trading. Allowance may be named differently in various carbon markets, such as EUA in the EU carbon market, or NZEA in New Zealand carbon market. But they are all in all under the cap and trade mechanism. The trading products in offset markets are emission reduction credit, which is usually supplements to allowance market, but may sometimes develop into voluntary emission reduction market. Emission reduction credits under various offset mechanism are also named differently, such as those certified by GS are called GSVERs, those certified by VCS are called VCUs. As shown in Fig. 5, the fundamentals of offset mechanism is to allow the emission reduction generated from GHG emission reduction projects (beyond carbon trading system) to offset GHG emitted by countries or companies undertaking emission reduction obligation, so that these countries or companies are provided with another way of accomplishing emission reduction obligations in addition to submitting emission allowance (Table 6).

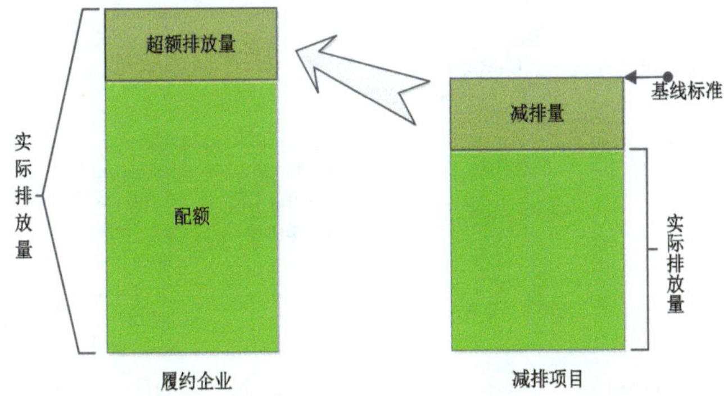

实际排放量	Actual emission
超额排放量	Emission beyond allowance
配额	Allowance
履约企业	Compliant companies
减排量	Emission reduction
极限标准	Extreme standard
减排项目	Emission reduction project
实际排放量	Actual emission

Fig. 5 Functional diagram of offset mechanism

Table 6 Categorization of carbon market

Types of carbon trading	Legal basis	Trading products	Motive of trade	Buyers	Sellers
Emission Trading (ET) (mandatory-allowance)	Kyoto Protocol	AAUs	Fulfilling the convention	Developed countries	Developed countries
Joint Implementation (JI) (offset)		ERUs	Mainly on fulfilling the convention	Developed countries, companies under the European carbon market	Developer of emission reduction projects in developed countries

(continued)

Table 6 (continued)

Types of carbon trading	Legal basis	Trading products	Motive of trade	Buyers	Sellers
Clean Development Mechanism (CDM) (offset)		CERs	Mainly on fulfilling the convention	Developed countries, companies under the European carbon market	Developer of emission reduction projects in developing countries
Domestic or local carbon market (EU, New Zealand, Korea, California, and seven China pilots) (mandatory-allowance)	Domestic or local laws (non-Kyoto)	Various AAUs	Fulfilling the convention	Carbon trading enlists common carbon trading participants such as companies and investment institutions	
Local offset mechanism (such as CCER in China) (offset)		Various ERUs	Mainly on fulfilling the convention	Carbon trading enlists common carbon trading participants such as companies and investment institutions	Developer where the offset projects are located
Voluntary offset mechanism (VCS, GS) (voluntary offset)	Self-initiated rules by civil groups (non-Kyoto)	Each ERUs	Mainly on voluntary emission reduction	Mainly companies undertaking CSRs	Developer where voluntary emission reduction projects are located

3 Status and Future of Carbon Market

3.1 Situation of International Carbon Market

The year 2005 is the starting year of the global carbon market. EU ETS also begin Phase 1 operation in this year. New Zealand, 9 states in Northeast US, California

State, Quebec of Canada, Republic of Korea, and many other countries and regions also began legislation to set up domestic or regional carbon market. By July 2017, there are 18 carbon trading markets around the world, covering 38 countries, 15 states (provinces) and 7 cities (Fig. 6).

Since 2005, the global carbon market used to be in full swing. Trade volume of the market in 2005 totaled US$11 billion, which reached US$176 billion in 2011, 16 times more than the original. Due to policy uncertainty during the negotiation of 2nd commitment period under the Kyoto Protocol and worldwide economic recession, trade volume and price in the EU carbon market faced major landslide since 2011. The market suffered a sharp decrease in trade volume and stood at US$30–50 billion between 2013 and 2016. Trade volume and trend of the global carbon market is shown in Fig. 7.

3.2 *Trend of Carbon Markets in Main Countries*

(I) **EU ETS**

EU ETS is a key policy approach for the EU to achieve emission reduction targets. It is started on January 1, 2005, with four implementation period so far.

The first period covers 2005–2007. It is a pilot period, with the main purpose of "learning from practice" and summing up the experience for future use. The EU had passed Directive 2003/87/EC for setting up emission trading scheme as early as June 2003, which is used as a basis to create the very first mandatory cross-border carbon emission trading mechanism in the world. EU ETS incorporates 12,000 emission sources from such sectors as energy, iron and steel, cement, glass, ceramics, and pulp-making, covering 45% of GHG emission in the EU. The EU emission allowance (EUA) is decided by member countries, with each submitting its own national allocation plan (NAP) to the European Commission. The European Commission will assess all NAPs and grant implementation after adoption.

The second period covers 2008–2012. This period coincides with the first commitment period of the Kyoto Protocol. It is a major policy approach for the EU to fulfill its targets under the Kyoto Protocol and related domestic targets. Starting from this period, EU ETS includes Iceland, Norway, and Liechtenstein, in addition to the 27 EU member states. This period assigns stricter emission allowance. EU tuned down the total allowance submitted by member states by 10.4%, and the penalty for exceeding emission limit was raised from 40 €/tCO_2 in the first period to 100 €/tCO_2.

The third period covers 2013–2020. The EU further enhances EU ETS by the revised Directive 2009/29/EC, which mainly changes the following: First, the coverage is wider. More sectors and GHGs are included. Second, control over emission limit and allowance are tighter gradually. Total emission limit will be directly set by the EU. Third, allowance allocation gradually transits to allowance auction. The auction will take the lead, with free assignment as a supplement. Fourth, the emission

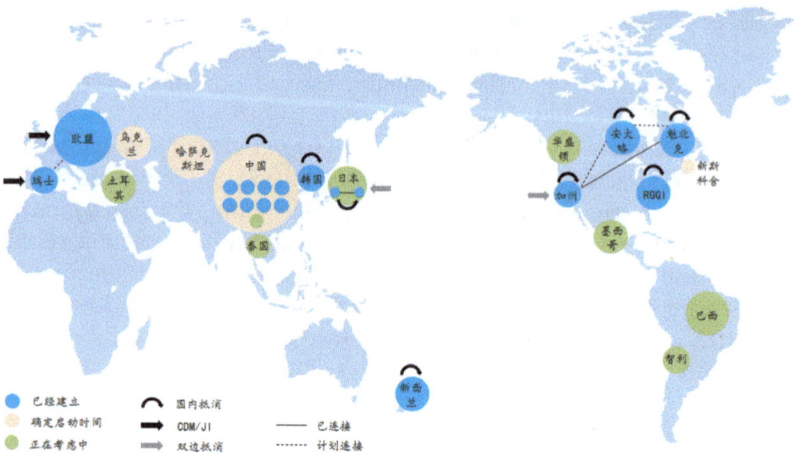

欧盟	EU
瑞士	Switzerland
乌克兰	Ukraine
土耳其	Turkey
哈萨克斯坦	Kazakhstan
中国	China
泰国	Thailand
韩国	Republic of Korea
日本	Japan
新西兰	New Zealand
华盛顿	Washington
加州	California
安大略	Ontario
魁北克	Quebec
墨西哥	Mexico
新斯科舍	Nova Scotia
巴西	Brazil
智利	Chile
已经建立	Already established
国内抵消	Domestic offset
确定启动时间	Confirming startup time
已连接	Already connected
正在考虑中	Considering
双边抵消	Bilateral offset
计划连接	Planning connection

Fig. 6 Distribution of global carbon trading system

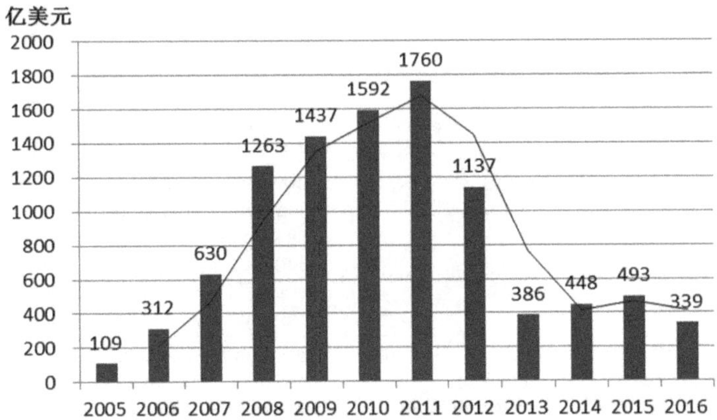

亿美元	US$100 million

Fig. 7 Historical trade volume and trend of global carbon market

reduction credit is further controlled. New CDM projects after 2012 will be accepted by the EU only if they originated from the least developing countries. The aviation sector is included since 2012. The emission allowance of various air companies will be calculated based on their carbon emission during 2004–2006 when flying in and out of the EU.

The fourth period covers 2021–2030. The European Parliament Committee on Environment, Public Health and Food Safety (ENVI) adopted the reform proposal of EU ETS after 2020. Based on ENVI, the European Parliament adopted Phase 4 reform plan of EU ETS on February 15, 2017. The Parliament agreed to most proposals relating to ENVI, but rejected some. First, the total allowance of EU ETS is decreasing by a linear reduction factor (LRF) of 1.74% annually. Though LRF may be raised to 2.2% in Phase 4, the European Parliament rejected the proposal to raise LRF to 2.4%. Second, the European Parliament refused the border adjustment measure (BAM) proposed by ENVI. BAM demand those importing cement from outside the EU to surrender their allowances. This proposal is intended to alleviate the compliance burden for the cement sector in the EU, so as to facilitate free allowance to shift towards allowance auction in this sector. Third, the European Parliament also propose to include the navigation sector, noting that CO_2 emission of ships arriving and departing and inside EU ports will be calculated from 2023. If the International Maritime Organization (IMO) still doesn't impose any measures in 2021, the European Parliament will propose to include EU and international navigation (covering non-EU ships) in the EU ETS.

Immediately after that, the Council of the European Union, on behalf of environmental authorities of member states, reached consensus on the Phase-4 reform of EU ETS on February 28, 2017. The table below compared the proposals adopted separately by the European Parliament and Council of the European Union (Table 7).

Table 7 Comparison of proposals by European Parliament and Council of the European Union

European Parliament	Council of the European Union
By 2021, 800 million allowances will be permanently eliminated from market stability reserve	If needed, allowances will be continuously eliminated from market stable reserve after 2023, instead of eliminating 800 million allowance at one time. At the beginning of each trading period, the allowance in the market stable reserve shall not exceed the auctioned amount of allowance of the last period. The exceeded part will be automatically annulled
Speed up the handling of extra carbon allowance in the market. Between 2019 and 2022, the allowance withdrawal rate of MSR will be up from 12 to 24%	Speed up the handling of extra carbon allowance in the market. Between 2019 and 2023, the recoverable allowance of MSR will be up from 12 to 24%
From 2021, LRF will be raised to 2.2% from the existing 1.74%, valid until 2024	Same as the resolution of European Parliament
Starting from 2023, CO_2 emission of ships arriving, departing and inside EU ports will be calculated	–
2% from allowance auction profit will be transacted to Modernization Fund to fund modern energy sectors in 10 low-income member states	Same as the resolution of European Parliament
For sectors with severe carbon leakage, such as iron and steel, cement, the allowance will be assigned 100% free of charge	Same as the resolution of European Parliament

Currently, the European Parliament and Council of the European Union have clear positions. The next step is Tripartite negotiation among the European Parliament, Council of the European Union and European Commission. It is expected the outcome will be reached in the second half of 2017 with regards to the phase 4 reform of EU ETS.

EUA trade is the absolute major global carbon market. Thus, EU carbon market is critical for the global carbon market, and it is also promoting the global carbon market to a certain extent.

Figure 8 shows the percentage of the trade volume of the EU carbon market in the global carbon market in recent years. In 2013, the trade volume of EUA was US$36 billion, accounting for 93.3% of that for the global carbon market. With the emerging carbon markets and the decreasing price of EUA, EU's market share begins to slide. The trade volume in 2016 was US$27.7 billion, and market share was down to 81.7%.

(II) **NZ ETS**

In November 2002, the New Zealand Parliament adopted climate change countermeasure act. In September 2008, GHG emission trading plan was incorporated into the act. In 2010, New Zealand successfully introduced domestic GHG emission

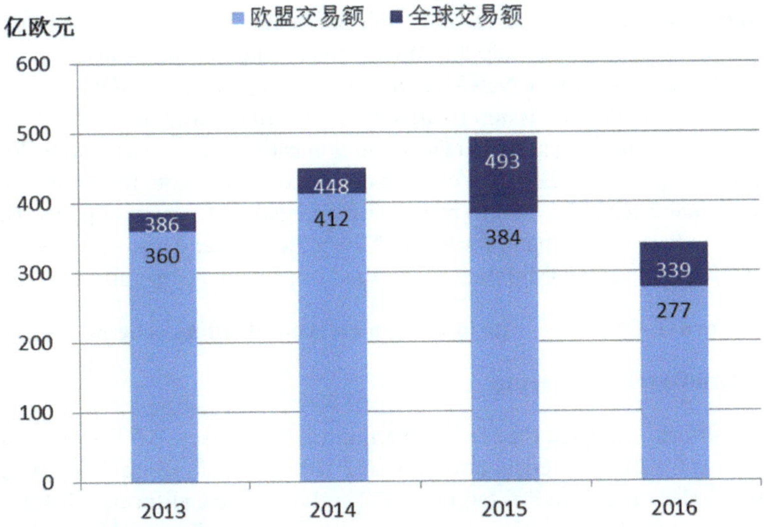

亿欧元	100 million Euro
欧盟交易额	EU trade volume
全球交易额	Global trade volume

Fig. 8 Percentage of EU ETS trade volume in global carbon market

trading scheme, making itself the second developed country (the first being the EU) to implement mandatory GHG cap and trade scheme.

The New Zealand Emission Trading Scheme (NZ ETS) covers a wide range, but adopts a gradual incorporation step for different sectors at different stages. Forestry projects were the first to be included in carbon trading system on January 1, 2008; followed by fixed energy, industry and liquefied fossil fuel on July 1, 2010; waste and synthesized gas in 2013; and agricultural sector (50% of total emission) in 2015.

Years 2010–2012 is the initial operating phase. In order to reduce the negative impact on the economy brought by the emission trading scheme, regulated companies are entitled to favorable policies at the initial operating phase. For every 2 tons of emission generated, only 1 ton of NZU need to be submitted, thus complying companies only need to pay 50% of compliance cost. Companies may purchase NZU freely from the market, or purchase from the government at the fixed rate of NZD 25/ton. In order to stabilize the carbon market, New Zealand adopted a piece of legislation in May 2016 to gradually eliminate the "2:1" payback obligation before January 1, 2019. Starting from 2017, complying companies will pay 67% of compliance cost, which will go up to 83% starting from 2018, and followed by 100% cost since January 2019. The detailed implementation plan is under discussion.

The NZ ETS encourages and certifies emission reduction credits for afforestation, so as to cut deforestation and help the country to accomplish emission reduction targets under the Kyoto Protocol. New Zealand allows companies to use CERs generated from CDM to meet their emission reduction targets. In order to avoid the inflowing global carbon credit to pressure down on its domestic carbon price, New Zealand sets limitation for using CERs. On one hand, this will contribute to maintaining its domestic demand for NZU, and to promoting domestic afforestation plans. But on the other, it reduces the demand for CDM. Given the smaller capacity of the carbon market, New Zealand's CDM policy has limited impact on the overall CDM market.

(III) California–Quebec–Ontario carbon emission trading scheme

1. California–Quebec scheme

In 2006, California adopted the Global Warming Solution Act (AB32), establishing the target that California's GHG emission in 2020 will be down up 20% compared to 1990. In order to achieve this target, AB32 authorize the California Air Resource Committee to formulate emission reduction measures, including market mechanism.

According to the plan, the emission trading scheme falls into three compliance period of 2013–2014, 2015–2017, and 2018–2020. On July 18, 2017, California Parliament, with the absolute support from both parties, adopted AB398 and AB617 acts with 2/3 majorities, meaning that the State's cap and trade plan will be extended until 2030. At the initial stage, the Scheme covers electricity, oil refining, cement, paper, glass, and 600 emission sources from 360 companies of annual emission over 25,000 tons. It also covers emission allowance totaling 165.8 million tons of CO_2e, and will be decreased by 2% annually. Starting from 2015, the coverage will be extended to natural gas, propane and traffic fuel. The covered sectoral emission totals 85% of the State, and the emission allowance will be 394.5 million tons of CO_2e, with an annual decrease of 3%. The allowance will be reduced to 334 million tons of CO_2e in 2020.

The California carbon trading mechanism sets the minimum carbon price at US$10/ton, and reserves 4.6% of the emission allowance in advance to avoid over-pricing in case of the necessary increase of supply. The mechanism also allows companies to accomplish 8% emission reduction obligation by offset mechanism, with the acceptable emission reduction projects covering forestry, urban greening, dairy CH_4, CFCs, and HCFCs refrigerant breakdown. But CERs generated from CDM projects are not accepted.

Quebec Cap and Trade system was legislated in 2009 and the initial ETS rules were created in 2011. The rules are further revised in December 2012, so that Quebec and California can establish friendly cooperation partnership. On January 1, 2014, California is linked to Quebec ETS, and both started the first joint allowance auction on November 25, 2014. This was the first time that cap and trade system are linked together between provinces (states) of different countries. California–Quebec carbon market linkage not only reduces carbon emission effectively, but also increases market flexibility, thus providing a good exemplary model for other carbon markets to initiate such linkage.

2. **Ontario scheme**

On February 25, 2016, Ontario released the *Management Regulations* (*draft*) on carbon emission trading scheme, which stipulates that the threshold is set at 25,000 tons of CO_2 equivalent per year, and those with annual emission over 10,000 CO_2 equivalent can choose to participate on a voluntary basis. In order to achieve the emission target of 37% reduction in 2030 in comparison to 1990, local government will distribute an initial maximum allowance of 142.3 million tons of CO_2 equivalent in 2017, which will decrease each year at the rate of 4.2% until the end of implementation phase I in 2020. The controlled organizations can use carbon offset credit to complete their 8% of their obligation. Manufacturing company, biomass power generation company, and those choose to participate may be granted free allowance. The remaining allowance will be auctioned quarterly. The minimum auction price in 2017 will be based on that occurred in California–Quebec carbon trading scheme, and will keep a similar 5% annual increase (excluding inflation factors). On May 18, 2016, Ontario officially adopted this regulation.

Ontario's emission trading scheme was officially started in early 2017, and it is intended to join the California–Quebec scheme the next year.

(IV) **Carbon trading mechanism in the Republic of Korea**

At the end of 2008, the Republic of Korea sets low-carbon green growth as a new development goal for the country. Korea adopted the Framework Act on Low-carbon, Green Growth in 2010, which stipulates that GHG quantity control and emission trading scheme shall be adopted.

On May 2, 2012, National Assembly of the Republic of Korea adopted the law to introduce a carbon trading mechanism. Republic of Korea's carbon trading mechanism will be started on January 1, 2015. It will cover more than 300 large-scale emission companies in power, iron and steel, petrochemical and pulp sectors (60% of total emission). This will contribute greatly to achieving the goal of 30% reduction of GHG emission in 2020 as compared to business as usual. At the initial stage, 95% emission allowance will be distributed to companies free of charge, the remaining will be arranged by means of auction.

Republic of Korea carbon trading scheme was officially started on January 12, 2015. According to Bloomberg New Energy Finance, between 2015 and 2020, the Republic of Korea's carbon market is expected to generate 130 million tons of emission reduction credits. Since the allowance was limited and domestic emission reduction credits were insufficient, the Republic of Korea's carbon trading law was revised accordingly in terms of global CER purchase at the end of March 2017. It allows the use of foreign emission reduction credit at implementation phase II (starting from 2018 namely), which is expected to give rise to the demand capacity of more CDM and globally offset projects.

3.3 Future of International Carbon Market

Currently, carbon trading keeps spreading forcefully around the world, with more countries and regions considering setting up a carbon trading market. The World Bank announced to initiate Partnership for Market Readiness (PMR) during the Cancun climate change summit in 2010, which will provide funding and technical support for developing countries, encourage and help them to set up market-based carbon emission reduction mechanism and organize a technical and experience-sharing workshop on a regular basis. China, Mexico, India, Indonesia, South Africa, Chile and other developing countries submitted to the WB their plans for developing carbon markets. They have obtained PMR funding support and initiated the preparation and development of domestic carbon markets. In addition, the World Bank also created the Carbon Pricing Leadership Coalition (CPLC), pooling together carbon market experts from governments, companies, universities, research institutes, and other stakeholders. CPLC regularly shares and promotes experience and case study in carbon market implementation, so as to promote and spread the carbon market globally.

Meanwhile, with the Kyoto Protocol coming to an end, the bond that links global carbon markets is disconnected, and global carbon market is thrown into a dispersed and fragmented development, with each country for itself. Such dismantled carbon markets come out without any unified rules and standards, which restricts the flow of capital and technology necessary for emission reduction, and discourages carbon market advantage in reducing emission reduction cost. But countries have never stopped their efforts in seeking a unified carbon market. On one hand, those with carbon markets have already made progress in cross-regional carbon market linkage, for example, EU has integrated carbon markets with Switzerland and Sweden, California and Quebec are also successfully connected. These practices offer a possible route for carbon market integration and unity. On the other, as an alternative of the Kyoto Protocol, the Paris Agreement also makes framework arrangement for the carbon market, which will have a great impact on how to cooperate on carbon trading and how global carbon market will evolve.

4 China's Policy on Combating Climate Change and Low-Carbon Transition

4.1 Necessity of Low-Carbon Transformation in China

For China, low-carbon green growth and achieving low-carbon transformation is the only way for sustainable social and economic growth and ecological civilization. Coping actively with climate change and developing low-carbon economy is not only the external demand for climate change, but also the internal requirement for energy conservation, emission reduction and breaking down the resource and environment constraints. It is also key measures for transforming economic development mode and seizing development advantage in the new round of "low-carbon revolution".

(I) **Urgent agenda of combating global climate change**

Currently, international community has reached consensus in the Paris Agreement regarding the target of controlling the average global temperature increase within 1.5 °C, as compared to 2 °C during the pre-industrial period. However, it is a daunting task for the world if we continue with the current development trend. China's GHG emission is large and increases rapidly. China is the largest emitter in the world and also the major source country of CO_2 emission increase. As the world is working to combat climate change and global carbon emission is under tighter control, there has arisen a sharp conflict between the energy consumption and rapid CO_2 emission increase in China's industrialization and urbanization on one hand, and global GHG reduction effort and climate protection objective on the other.

(II) **Low-carbon development is the fundamental approach to China's ecological civilization**

While China's economy enjoys rapid growth, it also pays a high price. The protruding conflict between social and economic growth and population, resource and the environment is increasingly becoming a bottleneck for development. In order to ensure sustainable economic and social development, the ecological civilization is put as a priority of national strategic agenda. President Xi Jinping made several major remarks with respect to ecological civilization, noting repeatedly that "lucid water and lush mountains are invaluable assets." In light of green and low-carbon development concept, we need to incorporate ecological civilization into economic, political, cultural, and social development and throughout these processes. Low-carbon development is the fundamental path and approach to promoting ecological civilization. As China's economic and social development move into the new normal, it is of critical importance that low-carbon development will contribute to ecological civilization and help to achieve economic low-carbon transition.

(III) **Low-carbon development is conducive to seizing economic development advantage for the future**

Against the background of combating climate change, low-carbon energy core technology and low-carbon economic growth reflect the core competitiveness of a country, a region or a company. The low-carbon industry is endowed with great market potential and new economic growth point, which is also a major bone of contention for the global economy, trade, and technology. China's industry should make plans as early as possible to take the opportunity to bring in new chances for industry upgrading and technology innovation. For the time being, the future decade will still be a golden period for China's peaceful development. Achieving the transition of development mode by oneself and green and low-carbon development in a flexible international context is key to moving from a country with economic strength to a country with economic power.

Whatever perspective, the only way for China is to pursue the green and low-carbon economic development path, promote technology innovation, develop low-carbon energy technology, and transform economic development mode and consumption mode. This is fundamental to coordinating economic growth and climate protection. Low-carbon economy and low-carbon transformation have become the trend of economic and social reform in the world. It is also the necessary path for China to cope with climate change in the framework of sustainable development.

All in all, low-carbon transition is a strategic option for China on the basis of current condition at home and abroad. It is also the only way out for China to combat climate change in the framework of sustainable development.

4.2 China's Regime and Mechanism for Climate Change

China has set up and is gradually improving the management regime and working mechanism for climate change. China sets up the National Leading Group to Address Climate Change. The management regime and working mechanism for combating climate change features unified leading group that was created and is being upgraded, centralized management by the National Development and Reform Commission (NDRC), responsibilities that are divided among various departments and wide participation of all regions and sectors.

In June 2007, in order to strengthen the leadership of coping with climate change and energy conservation and emission reduction, China set up the leading group of address climate change and energy conservation and emission reduction. The group is also known as the National Leading Group to Address Climate Change or the Office for the Leading Group of Energy Conservation and Emission Reduction of the State Council in case of need (one body with two different titles). Headed by the Premier of the State Council, the Group has as its member the senior officials of more than 20 departments including NDRC and Ministry of Foreign Affairs. The NDRC serves as the coordinating body in combating climate change and energy conservation and emission reduction, and undertakes practical work of the Leading Group.

The main job of the Leading Group covers the following: drafting national strategy, policy and countermeasure in coping with climate change, implement the work arrangement for combating climate change in a unified manner, review the proposal for international cooperation and negotiation, and coordinate to resolve major issues in climate change. Upon formation, the group reviewed and adopted *China's National Plan for Combating Climate Change* and *Comprehensive Work Plan for Conserving Energy and Reducing Emissions*. Hence, China becomes the first developing country to draft and implement the national plan for combating climate change.

The group further creates the office for the National Leading Group to Address Climate Change and the office for the Leading Group of Energy Conservation and Emission Reduction of the State Council, both of which are inside the NDRC and undertakes daily work of the Leading Group. The National Leading Group to Address

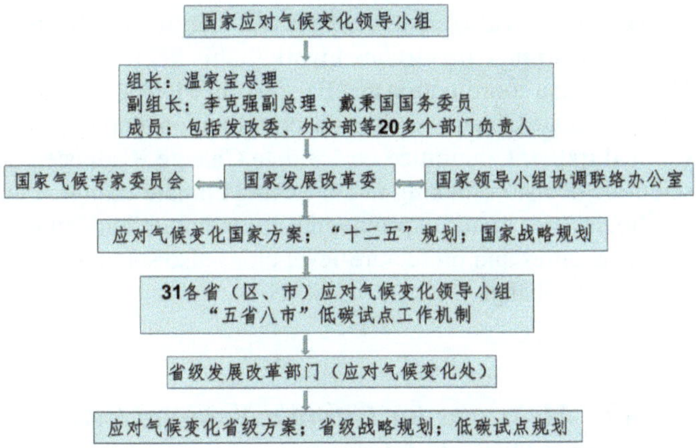

国家应对气候变化领导小组	National Leading Group to Address Climate Change
组长：温家宝总理	Group leader: premier Wen Jiabao
副组长：李克强副总理、戴秉国国务委员	Deputy group leader: Vice Premier Li Keqiang, State Counselor Dai Bingguo
成员：包括发改委、外交部等 20 多个部门负责人	Member: Leaders of more than 20 departments, such as NDRC, MOFA
国家气候专家委员会	National Climate Experts Committee
国家发展改革委	NDRC
国家领导小组协调联络办公室	Coordination and liaison office of the leading group
应对气候变化国家方案；“十二五”规划；国家战略规划	National plan for combating climate change; 12th Five-year Plan; national strategic planning
31 个省（区、市）应对气候变化领导小组	Leading groups to address climate change in 31 provinces (autonomous regions, cities)
“五省八市”低碳试点工作机制	Low-carbon pilot mechanism in five provinces and eight cities
省级发展改革部门（应对气候变化处）	Provincial development and reform departments (climate change divisions)
应对气候变化省级方案；省级战略规划；低碳试点规划	Provincial plans for combating climate change; provincial strategic planning; low-carbon pilot project plan

Fig. 9 Organizational chart of regime and mechanism for national plan for combating climate change (in year 2010)

Climate Change is improved and enhanced on the basis of the national coordination group for climate change countermeasures. For the Office for the Leading Group of Energy Conservation and Emission Reduction of the State Council, its coordination and energy conservation function is delegated to NDRC, while pollution control function falls under the formerly SEPA. China's framework for combating climate change is shown in Fig. 9.

Upon the formation of the group, NDRC officially set up the Department of Climate Change in 2008, which is responsible for coordinating and managing climate change. Other organizations also created bodies and working mechanism for climate change, in an effort to arrange their related work in this field.

In 2010, China set up a coordination and liaison office inside the National Leading Group to Address Climate Change to enhance coordination among various organizations. The National Expert Committee on Climate Change is adjusted and enriched to enhance scientific decision-making in combating climate change. Leading group and specialized bodies are created at provincial, autonomous, and municipality level. Some sub-provincial cities and prefecture-level cities also set up their own working mechanism for climate change.

4.3 *Main Policy Measures for China's National Plan for Combating Climate Change*

The international community has furthered its understanding and consensus in low-carbon development goal and global governance mechanism. Carbon trading mechanism is increasingly being accepted as a major policy tool for carbon emission reduction by major countries. Against this background, the Chinese government also put low-carbon development at an unprecedentedly strategic high level, and it has become the core principle for guiding China's future economic and social development.

In August 2009, the National People's Congress (NPC) adopted the Resolution of the Standing Committee of the National People's Congress on Making Active Responses to Climate Change, clearly stating that combating climate change in an active approach shall be incorporated in national economic and social development plan for achieving long-term task of sustainable strategy. The National Climate Change Act is also under discussion. The 3rd plenary session of the 18th National Congress of the CPC officially noted that ecological civilization shall be enlisted in the development system to seek economic, political, cultural, social, and ecological progress, with the focus on low-carbon and green economic growth. NDRC and other ministries are working to draft National Climate Change Program and studying emission peak. They not only study China's mid and long-term emission reduction plan, but also make preparation for low-carbon and green transformation in the future.

In November 2014, China and US presidents issued the Sino-US Joint Declaration on Climate Change in Beijing, in which China states that its CO_2 emission will reach the peak by 2030 and it will try to top that as early as possible. China also plans to increase the use of non-fossil fuel energy by 20% in its primary energy mix by 2030. This is the first time for China to openly commit its goal in emission peak. In September 2015, China and US reiterate the Sino-US Joint Declaration on Climate Change, noting that both countries will enhance domestic emission reduction and international emission reduction effort, and will jointly advance the Paris climate change summit to an ambitious outcome.

In 2015, China submitted its INDC to the Secretariat of the UNFCCC, which states China's INDC goals by 2030 as follows: CO_2 emission will reach the peak by 2030 and will try to top that as early as possible; CO_2 emission per GDP unit will be down by 60–65% compared to 2005, non-fossil fuel will account for 20% in primary energy mix. Forest reserve will increase by 4.5 billion cubic meters compared with 2005. The paper also released China's low-carbon initiative and policy measures in coping with climate change, announcing to promote the trading market of carbon emission permits, and giving full play to market role in resource allocation.

In April 2015, the *Opinions on Accelerating Ecological Progress* put forward the following points: actively cope with climate change; effectively control GHG emission; practically promote low-carbon pilot projects in provinces, cities, town, industrial parks, and communities; constructively participate in the negotiation of climate change; and formulate a fair and sound framework for combating climate change. In September 2015, the *Overarching Plan for Reforming Ecological Civilization* lays out the following: gradually set up national regime for carbon emission quantity control and breakdown implementation mechanism, create effective mechanism to increase the carbon sink in forest, grassland, wetland, and ocean, enhance international cooperation in combating climate change.

In March 2016, *the Outline for the 13th Five-Year Plan for Economic and Social Development of the People's Republic of China* notes that China attached equal importance to adaptation and mitigation in combating climate change, effectively control GHG emission, implement the commitment to emission reduction, proactively adapt to climate change, enhance adaptability to climate change, expand international cooperation, participate deeply in global climate governance, and make contribution to combating climate change. Low-carbon development has become a key and necessary approach to resource conservation, environmental protection, and sustainable economic development. It is also the main task for the Chinese government during the 13th Five-year Plan.

In order to further guide China's low-carbon development and formulate an overarching design, the NDRC, with the approval from the State Council, released the *National Climate Change Program (2014–2020)*. Combining the plan and other documents from the central government, China's low-carbon development for the next phase can be summed up in four points:

1. Gradually, set up a regime for controlling carbon emission. China will make it a major target of setting the carbon emission control during the 13th Five-year plan period on the base of that during the 12th Five-year plan period. Starting with the carbon emission in the energy sector, a regime for controlling carbon emission and implementation assignment regime will be gradually put into place, which will cover the carbon emission and non-CO_2-type GHG emission generated from industry sector. By controlling the quantity, the economic development mode will be forced to change, and economic structure and energy mix will be prompted to make a strategic adjustment.
2. Set up a regime for carbon emission permit. On the basis of quantity control, key sectors and companies will be selected to implement carbon emission permits,

so as to identify sector emission standard, work out distribution plan of carbon emission permits, and implement an operating and supervision mechanism, where only those with emission permits are allowed to discharge.

3. Set up a trading and transfer regime for carbon emission permits. On the basis of emission permits, in order to achieve effective allocation of the limited resource of carbon emission reduction across the country, the emission permits should be allowed to be transferred and traded among companies, so as to foster a trading market.
4. Formulate detailed climate change policies in light of the functions of an area and sectoral characteristics. China is endowed with wide landscapes and diversified natural conditions. There exists imbalanced development among different regions and these regions have various main functions to play in China's economic and social development. Thus, in addition to nationwide policies, specific climate change policies shall be implemented for different regions in accordance with actual conditions. Meanwhile, since carbon emission permit itself can't cover certain sector and areas (such as agricultural forest zone) due to intrinsic reasons, sector-specific policies have to be formulated in this regard.

4.4 Work Plan for Controlling GHG Emission During the 13th Five-Year Plan

On October 27, 2016, in order to advance green and low-carbon development, accomplish the low-carbon development goal identified in the 13th Five-year plan, and promote China's CO_2 emission to reach its peak around 2030 and early accomplishment, the State Council released the Work Plan for Controlling GHG Emission during the 13th Five-year Plan (Guo Fa (2016) No. 61) (hereinafter "the Plan"). The core objective of the Plan is to cut CO_2 emission per GDP unit by 18% in 2020 as compared to 2015, so that carbon emission can be effectively controlled. The Plan lays out comprehensive assignments and measures for low-carbon emission in major sectors such as energy system, industrial economy, and urban and rural development.

In the energy sector, it demands to initiate an energy revolution with low-carbon development. In terms of improving energy efficiency, it is clearly stated that dual control of energy consumption and energy intensity will be implemented. While requiring that the energy consumption per GDP unit in 2020 will be down by 15% as compared to 2015, the total energy consumption will be controlled within 5 billion tons of standard coal. This will foster the energy pattern where newly increased energy demand with be met by low-carbon energy. While controlling energy consumption quantity and reducing energy consumption increase, the Plan also raised clear requirement on optimizing the energy mix, making work arrangement by focusing on "controlling coal consumption, increasing gas supply and developing non fossil fuel energy." It demands to control total coal consumption and make coal consumption reach its peak as early as possible. By 2020, natural gas will account for 10% of total energy consumption, and that for non-fossil fuel energy will be raised to 15%. Meanwhile, it is clearly stated that by 2020, the CO_2 emission for power generation at large-scale power companies shall be controlled within 550 g of CO_2/KWh. Carbon emission efficiency will be adopted to promote green and low-carbon transition in power sector.

In the industry sector, it demands to create low-carbon industry system. For the first time, low-carbon development is regarded as the driving force for quality and efficiency economy, with an eye to promoting transformation and upgrade of industry structure. It notes that the control of carbon emission in the industry sector will be enhanced, efforts will be made to advance certain heavy and chemical industry reach their emission peak by 2020, CO_2 emission in industry production sector will be stabilized, and the CO_2 emission of per unit industry added value will be cut by 22% as compared to 2015, thus laying a solid foundation for the 2030 Objectives. The Plan also puts forward the requirement for the company. Emitting companies need to enhance carbon emission management, promotes new low-carbon technology and process, and adopts low-carbon benchmarking program to encourage key companies to start emission benchmarking initiative. By 2020, the carbon emission per unit for major energy-consuming products shall reach the international advanced level.

In the urban and rural sector, it demands to promote low-carbon development during urbanization. It is clearly stated that low-carbon concept and requirement should be implemented in urban and rural planning. Effort will be made to explore a new type of urbanization mode featuring intensive, intelligent, green and low-carbon development. Cities are encouraged to draft their low-carbon development plan. By 2020, the urban green building should account for 50% of new buildings. The public building should enhance low-carbon operation and management. The rural area should shift their energy consumption toward clean and low-carbon utilization, with pilot building projects of zero carbon emission. It is demanded to promote modern and comprehensive transportation system, develop low-carbon logistics, improve urban transport system with public vehicles as a priority and further implement low-carbon transportation pilots. Carbon emission standards for a new vehicle is being studies for the first time, adopting "carbon emission per 100 km" to advance green and low-carbon transition of the auto sector. By 2020, the CO_2 emission per turnover volume for the operating lobby, passenger bus and ship will be down by 8, 2.6, and 7% as compared to 2015. The CO_2 emission per passenger volume for the urban public bus will be down by 12.5% as compared to 2015.

In region development, the Plan categorizes reduction goals for carbon intensity at provincial level during the 13th Five-year Plan period, namely, 20.5% reduction for Beijing, Tianjin, Hebei, Shanghai, Jiangsu, Zhejiang, Shandong and Guangdong; 19.5% reduction for Fujian, Jiangxi, Henan, Hubei, Chongqing and Sichuan; 18% reduction for Shanxi, Liaoning, Jilin, Anhui, Hunan, Guizhou, Yunnan and Shaanxi; 17% reduction for Inner Mongolia, Heilongjiang, Guangxi, Gansu and Ningxia; 12% reduction for Hainan, Tibet, Qinghai and Xinjiang. The Plan also notes that support will be given to prioritize development zones to reach their emission peak. Other regions meeting the conditions are also encouraged to put forward their peak targets and approach. Efforts will be made to explore carbon emission control in some developed provinces and cities, and measures are launched to support low-carbon development in poor areas.

5 Prospect of China's Carbon Market

5.1 China's Carbon Market

In November 2011, the General Office of NDRC released the *Notice on Pilot Projects of Carbon Emission Trading*, which officially grants pilot projects of trading carbon emission permits in Beijing, Tianjin, Shanghai, Chongqing, Hubei, Guangdong, and Shenzhen. It is intended to achieve the action target of controlling GHG emission at a lower cost by using market mechanism, so as to accelerate the transformation of economic development mode and upgrade of industry structure. In December 2016, Fujian Province initiated a trading market for carbon emission within the province. Local carbon market generally operates as expected, which accumulates valuable experience for national carbon market and prompts the decision makers' determination in setting up national carbon trading market in a faster approach. In 2016, President Xi Jinping clearly announced that China will start the national carbon trading market in 2017.

(I) **Master plan for national carbon market**

National carbon market is a nationwide market with a unified standard. It trades in two major products (allowance and CCER) that can circulate freely across the country. According to NDRC's plan, national carbon market will be divided into three phases. Phase 1 (2014–2016) is the preparatory phase, and the system development is expected to finish by the end of 2016. Phase 2 (2017–2019) is the trial operation phase, with the main target of accumulating experience through trial operation and improve the system design. Phase 3 (post 2019) is the full implementation phase. On the basis of a complete carbon trading market, the coverage will be expanded, the trading rules will be improved, and derivative trade will be initiated (Fig. 10).

(II) **Framework of national carbon market**

1. **Legal system**

Establish "1+3" legal system: guided by Regulation on the Administration of Carbon Emissions Trading, the national carbon market will find its legal basis in *Management Decree on Emission Reporting and Verification*, *Management Decree on the*

	Study and development of all sectors related to national carbon trading system	Trial operation and gradual improvement	Full implementation
	2014ó2016	2017ó2019	Post 2019
Central	· Legislation, drafting management documents and technical standard · Reporting and verification of historical data · Start initial allowance allocation · Enhance infrastructure and capacity building	· Phase 1 · Trade of carbon emission permits · Market regulation and management	· Phase 2 · Expand coverage scope · Improve system and regulations · Study global linkage
Local	· Capacity building activities · Identify list of major companies and organize reporting and verification of historical emission · Allocate allowance according a unified approach	· Allocate allowance annually · Organize emission report, verification and compliance annually · Regime improvement and innovation in light of local condition	
Company	· Participate in capacity building activities · Set up internal reporting regime for carbon emission accounting · Fulfill reporting obligation and cooperate with support verification · Other companies may participate in carbon market through CCER	· Fulfill emission reporting and compliance responsibility · Improve carbon capital investment and management regime · Voluntary participation in the trade to reduce emission cost · Actively participate in political regime improvement and innovation	

Fig. 10 Roadmap of national carbon market development

Accreditation of Third-Party Verifiers and *Management Decree of Trading Activities*. The Regulation is being led by the Legislative Affairs Office of the State Council and under legislative review following the legislation review process.

2. **Scope**

National carbon market will gradually incorporate any companies in 8 major industries that have overall energy consumption over 10,000 tons of standard coal in any given year from 2013 to 2015. These eight industries are power, petrochemical, chemistry, construction material, iron and steel, non-ferrous metal, papermaking, and civil aviation.

3. **Monitoring, reporting and verification system**

Regarding data monitoring and reporting, it is applicable to use the GHG emission caculation method and reporting guidance for 24 major industries released by the NDRC. For third-party verification rules, it is applicable to use the third-party verification guide and the reference conditions for third-party verification body and personnel, both of which are released by the NDRC. Provinces have finished data verification of historical carbon emission during 2013–2015.

4. **Allowance allocation**

The initial allowance allocation in the national carbon market is mainly allocated free of charge. In the future, more focus will be on paid allocation, with main methods of baseline and historical intensity reduction.

Benchmark method: allowance = industry benchmark × product output volume

Historical intensity reduction: allowance = historical intensity × emission reduction coefficient × product output volume

The State Council approved the *Cap Setting and Allowance Allocation Plan of National Carbon Market* in 2016. The NDRC started test calculation of allowances in power, cement, and electrolyte aluminum in Sichuan and Jiangsu in May 2017.

5. **Offset mechanism**

On June 13, 2012, the NDRC released the *Provisional Management Method for Trade in Voluntary GHG Emission Reduction*, which identifies the application scope, managing authority, participants, recording keeping system, national registry for GHG emission reduction, and clarifies the managing rules governing voluntary emission reduction projects, emission reduction credit, emission reduction credits trade, review and verification processes. In January 2015, China's national registration system for voluntary emission reduction went online. There are 9 exchange agencies that are capable of starting CCER trade business across China, including 7 pilot exchange organizations Sichuan United Environment Exchange and Fujian Haixia Equity Exchange.

Currently, the NDRC pauses CCER projects and application for credits application, so as to further improve and revise the *Provisional Management Method for Trade in Voluntary GHG Emission Reduction*, optimize the trade market of voluntary emission reduction and study the offset rules for the national carbon market.

6. **Compliance rules**

The governments will assess companies' compliance obligation. If a company provides to government its allowance that is no less than the actual emission before the expiry date, then it is considered as conformity. The extra allowance can be traded or stored for future implementation. If a company submits its allowance that is less than the actual emission, then it is considered nonconformity. The company may choose either to buy sufficient allowance from the market to fill in the gap, or be fined at 3 times the average market price of the allowance one day before the expiry date.

(III) **Management system of national carbon market**

The national carbon market follows a tiered management of "central–provincial–local" approach. The NDCR serves as the managing authority of national carbon trading, with a focus on macro management. Its responsibilities include drafting national carbon market rules and standards, managing national registration system and trading agencies, managing the qualification of verification agencies and executing market regulation; provincial development and reform commissions focus more on micromanagement. Its responsibilities include specific execution and management of carbon trading activities within their administrative regions, which mainly covers data reporting and verification, allowance allocation, conformity assessment, regulation, and penalty. Local development and reform commissions work with their provincial counterparts to organize companies to submit their data report and fulfill their responsibilities (Table 8).

Table 8 Labor division between central and local governments in relation to the national carbon market

	NDRC	Provincial development and reform commission	Local development and reform commission
Monitoring reporting and verification	Identify technical standards and qualification requirement, responsible for managing the qualifications of verification agencies	Responsible for managing the monitoring, reporting, and verification of major emission organizations, as well as managing verification agencies	Implement the monitoring plan, urge companies to finish data reporting and verification
Coverage	Identify the standard for incorporate	Identify company list at the provincial level according to standards and have it submitted to the NDRC. The provincial coverage may be further expanded in the future	Identify company list at their administrative level according to standards and have it submitted to provincial development and reform commission
Allowance volume (Cap)	Identify the total quantity of national and local allowance allocation	–	–
Allowance allocation	Identify the allocation method and standard	Implement allocation according to standards in a strict manner and may start the paid allocation	Cooperate to start allowance allocation

(continued)

Table 8 (continued)

	NDRC	Provincial development and reform commission	Local development and reform commission
Allowance surrender (Compliance)	Publicize compliance status, draft penalty rules for noncompliance	Responsible for managing the enlisted companies to meet the requirements and execute the penalty	Urge companies to meet the requirement and assist in carrying out penalty
Registration system	Responsible for setting up and managing the system	Use provincial administrator's account to manage the allowance allocation and cancelation within the administrative jurisdiction	–
Allowance trade	Identify the trading agency	Supervise the trade inside provinces	–
Capacity building	Training instructor to help with local capacity building	Organize provincial companies to participate in capacity building	Organize companies under their administration to participate in capacity building

5.2 Challenges Facing China's Carbon Market

Currently, the specific regime, element design, technical standard, hardware facility for national carbon market are controllable, but they need to be further coordinated at the central government level, so as to unify national effort and foster sound interaction between the central and the local. The country will actively promote and provide support for capacity building, and the local and companies will also actively participate in the development of a national carbon market. But there are still challenges facing national carbon market development, including:

First, the legislation process needs to be accelerated. In December 2014, on the basis of releasing the Provisional Method for the Administration of Carbon Emission Trading (as a departmental rule of the NDRC), the NDCR submitted the *Regulation on the Administration of Carbon Emission Trading* by the end of 2015. The NDRC also finished several rounds of opinions collection from local levels and related departments. This has been listed as a legislation task for the State Council in 2016. In order to strengthen the support for national trade in carbon emission rules, it is needed to accelerate the release of the *regulation* and its related implementation rules.

Second, there are some difficulties in linking the pilot projects. Different pilot systems and regimes are designed with diversification, and allowances circulated in different pilot systems have various qualities, thus it is challenging to connect these pilots with a national carbon market. The work plan or proposal for effective connection between pilots and national carbon market is still underway. It is still unclear as to how the pilot regions will effectively connect with the national carbon market.

Third, the progress is not at the same pace for pilot regions to participate in the national carbon market. With the motivation, training and capacity building provided by the country, the provincial authorities at the pilot regions have reached consensus on developing a national carbon market. Some provinces like Fujian, Sichuan, Hebei, and Shanxi not only follow the central demand to start data reporting and verification of historical emission, but also actively initiate provincial preparatory work of carbon market, with efforts made in data reporting, verification of historical emission and capacity building of third-party verification agency. Other nonpilot regions progress slowly in developing carbon markets.

Fourth, it is needed to raise companies' awareness and capacity in participating in the carbon market. The industry associations and companies in nonpilot regions have attended several training in the carbon market, which further raised their awareness and capacity in effective participation in the carbon market. Generally speaking, large state-owned enterprises have a better acceptance of carbon market development, given their advantages in political understanding and knowledge, as well as the experience of key emission control companies in pilot regions. While private companies have less awareness and capacity in carbon market given political factors or insufficient management systems. In order to motivate these companies to participate, on one hand, it is needed to enhance political mobilization; on the other hand, it is needed to strengthen capacity building and technical guidance.

5.3 *The Impact of China's Carbon Market on Forest Carbon*

Forest carbon plays an irreplaceable role in mitigating global climate change and improving eco-environment. It is a key component of voluntary emission reduction market. However, the validity and permanence of such forestry projects usually result in little emission reduction quantity and have a complicated methodology. They are easily questioned and relatively not high on cost-effectiveness. Therefore, forest projects are not as competitive as conventional energy projects. By September 30, 2017, according to China's information platform on voluntary emission reduction trade, the quantity of emission reduction on record totals 287, with 254 put online. The total recorded quantity of emission reduction is 52.83 million tons of CO_2e, but forestry projects accounts for less than 1%.

However, forestry projects come with a special implication in improving eco-environment, protecting climate, and helping to alleviate poverty, which is not provided by conventional projects. As long as policy support is provided, these projects will enjoy a broader vision. Currently, there are two utilizations for China's forestry projects: one is for the offset products in obligatory carbon market; the other is for the carbon-neutral products in voluntary emission reduction trade.

(I) **Forest carbon for mandatory carbon market**

China's mandatory forestry market provides new space and opportunity for using forest carbon as offset products. At the state level, with the methodologies for forestry emission reduction projects approved by the government, it can become potentially qualified projects under the framework of *Provisional Management Methods for Trade in Voluntary GHG Emission Reduction*. At the local level, Guangdong and Fujian draw upon their advantage in abundant forestry resource to actively study the methodologies for provincial forest carbon and develop emission reduction projects for providing offset products to carbon markets within their administrations. Guangdong released the *Inclusive Methodology for Forest Protection in Guangdong Province*, and the *Inclusive Methodology for Forest Management in Guangdong Province*. Under the guidance of these two documents, the provincial inclusive quantity of emission reduction is kept on record, and the competitive bidding system is introduced for the certified provincial inclusive emission reduction. Fujian released the *Pilot Plan for the Trade of Forest carbon in Fujian Province*, which calls for reaching 2 million *mu* of forest area for forest carbon and increasing 1 million tons of new carbon sink during the 13th Five-year Plan period. The certified forest carbon project will be used to provide carbon offset products to carbon market in Fujian.

(II) **Forest carbon for voluntary emission reduction market**

In addition to offset products in mandatory carbon market, forest carbon can also be used in the trade in voluntary emission reduction and serve as supplementary projects for carbon neutrality. Carbon neutrality refers to calculating the CO_2 emission that is directly or indirectly generated by companies, organizations or individuals in a certain period. By purchasing the quantity of carbon emission reduction credits or increasing carbon sink through afforestation, the emitted CO_2 is offset, thus achieving the goal of zero carbon emission (Table 9).

On August 22, 2017, the BRICS Xiamen Summit on carbon-neutral project is started in Xiamen. The summit meeting adopting the approach of absorbing the CO_2 generated during the summit by afforestation, thus achieving zero emission target. This is the first of its kind for senior leaders of BRICS countries. It is also the 4th large-scale international conference organized by the Chinese government by means of afforestation, following the United Nations Climate Change Conference in Tianjin in 2010, the Beijing Summit of APEC Leaders in 2014 and the G20 Hangzhou Summit in 2016.

Generally speaking, there are not many voluntary emission reduction activities in China. But with the continuous improvement of voluntary emission reduction market in China and the concerted efforts of all parties in this market, it is believed that, in the foreseeable future, more companies and individuals will participate in the voluntary emission and achieve sustainable development path through scientific management of carbon emission.

Table 9 Participants and purpose of voluntary emission reduction market

Participant	Purpose
Organizer of large-scale events	Offset the carbon emission generated from events, build up social image of low-carbon and green development
Enterprise	For the perspective of social responsibility, purchase certain amount of voluntary emission reduction indicator to achieve carbon neutral; participate in the trade and operation of emission reduction indicators
Individual	Actively participate in carbon emission reduction and achieve carbon-neutral individual
Financial institution, environmental exchange institution	Develop matching financial products, provide all-round financial service at tier-2 market trade in terms of project financing and emission reduction credit

Response to Climate Change by China's Forestry and Vision of Forest Carbon Market

Chunfeng Wang

Summary

This chapter summarized the international forestry-related negotiation process under the United Nations' Framework Convention on Climate Change (UNFCCC) and its Kyoto Protocol (KP). The negotiation results of the carbon accounting rule for Land Use, Land Use Change of Forestry (LULUCF), and Reduce Emission from Deforestation and Forest Degradation in developing countries are also included in this chapter. We try to analyze China's main forest policy for combating climate change, to identify the opportunities and challenges faced by China's forestry sector under the framework of Climate Change. We also introduce China's potential forest carbon increment and the approach to maintain the forest carbon increment. Based on the theory of carbon trading scheme and current development pathway, we predict the international and China's forest carbon market and provide the suggestions for China's policy of forest-related climate actions and forest carbon markets.

Forestry sector is playing an important role in combating climate change. The international forest-relevant climate negotiation includes policy and technique issues and mainly focuses on technique issues. The methodology of forest carbon accounting under Kyoto Protocol is still very complicated, and still needs more negotiation and more technical support.

The policies to combat climate change are offering new opportunities for China's Forestry sector, with the increasing attention from government and public, more and more people realized the importance of forest for combating and adapting to climate change.

Since 2003, China has formed its forestry policy to address the issue of Climate Change, however, it still has big space to improve the implementation of the policy. The large scale of reforestation in the past 30 years has increased China's forest carbon and it predicted that China's forest carbon can neutralize 7% of China's

C. Wang (✉)
State Foretry and Grassland Administration of China, Beijing, China
e-mail: chunfengw@126.com

Z. Lu et al. (eds.), *Forest Carbon Practices and Low Carbon Development in China*, https://doi.org/10.1007/978-981-13-7364-0_2

emission from manufactural industry. China still has big potential to increase its forest carbon reduction through the measures of afforestation, reforestation, sustainable forest management, forest and wetland conservation as well as the promotion of biomass and good use of wood products.

Although the carbon trading scheme can reduce the cost of carbon reduction, it is still controversial and the demand from the developed countries is still very weak, we predict that the global forest carbon market will still be unmatured until 2020. China is developing its domestic carbon market and will offer opportunities for China's Forest carbon credit. China's forest policymakers should try their best to achieve China's goal for international commitment and also should lead the design of China's Forest carbon market. China still needs to link the forest carbon with other industrial sectors, establish its forest carbon monitoring system, and integrate the forest carbon into the entire national carbon trade scheme.

1 Negotiations on Forest Issues Under the UNFCCC and Kyoto Protocol

(I) Origin of negotiation on forestry issues

Before the UNFCCC and Kyoto Protocol came into being, scientific community already had high consensus on the close relationship between forestry and mitigation and adaptation. After the UNFCCC becomes effective, the Contracting Parties are committed to using comparable approach under Article 4.1 for compiling and issuing all anthropogenic GHG emission and absorption list. With sustainable management, carbon sink is further enhanced in land-based ecosystem including forest. In order to achieve GHG control target set in the UNFCCC, the Kyoto Protocol was adopted in 1997, with Articles 3.3, 3.4, 3.7, 6, 12 touch (namely industrial countries and countries transiting to market economy, hereinafter "developed countries") upon specific issues on how forestry contributes to Annex I countries in quantifying their commitment to emission reduction. Circling around there articles, a series of negotiations on specific forestry issues were conducted under the UNFCCC and the Kyoto Protocol.

Under the UNFCCC, negotiation issues that are related to forestry before 2005 mainly focus on how to estimate and report land use, land use change, and forestry-caused carbon source/sink change in national GHG list. Negotiation is organized by the Intergovernmental Panel on Climate Change (hereinafter "IPCC"), which is authorized by the Conference of Parties (COP) to the UNFCCC to draft guidance on national GHG report list. Discussion is mainly focused on how IPCC guidance can be used for GHG report and reporting format. As IPCC continues to update and improves on national GHG list reporting, negotiations in this regard have been continuing and will keep on continuing in the future.

During the COP11 of UNFCCC held in Montreal, Canada at the end of 2005, parties accepted a joint proposal submitted to the Secretariat by Papua New Guinea and

Costa Rica on "reducing emission from deforestation in developing countries: principles of incentive action", which leads to negotiations on how to involve forestry in developing countries in combating global climate change against the background of the UNFCCC. During the COP13 held in Bali, Indonesia in 2007, parties agreed that developed countries will adopt effective incentives mechanism and policy measures to encourage parties from developing countries to reduce deforestation and emission resulted from forest degradation, and promote forest conservation, sustainable management of forest, and enhancement of forest carbon stocks (REDD+), which are covered in the negotiations on the "Bali Roadmap". Since then, REDD+ becomes a major forestry negotiation issue under the UNFCCC.

Under the Kyoto Protocol, since the Kyoto Protocol came into being at the end of 1997, countries are engaged in a series of negotiations on how to utilize carbon source/sink caused by land use, land use change and forestry to achieve quantifying emission reduction committed by developed countries during 2008–2012 (namely 1st commitment period). The negotiation topic is called "land use, land use change and forestry (hereinafter LULUCF)", which mainly focuses on how to calculate and report carbon source/sink change of LULUCF during the 1st commitment period that covers afforestation, reforestation, deforestation, forest management, farmland management, pasture management, and vegetation recovery in developed countries since 1990, and how to connect these changing calculation outcomes with emission reduction targets during the 1st commitment period for developed countries. Negotiation on LULUCF during the 1st commitment period was generally finished by the end of 2001, and the consensus reached was included in the package of "Marrakech Agreement". The "Marrakech Agreement" provides specific rules to follow for fully implementing the Kyoto Protocol articles. After that, the COP11 by the end of 2005 launched negotiation on the 2nd commitment period under the Kyoto Protocol, namely how developed counties will continue the negotiation on quantifying emission reduction by referencing the Kyoto Protocol mode after 2012, also known as "negotiation on the 2nd commitment period of the Kyoto Protocol". During this round of negotiation, developed countries note that there are numerous problems in LULUCF during the 1st commitment period, which is not conducive to giving LULUCF's role in mitigating climate change. They demand to modify the rules for LULUCF during the 1st commitment period. Therefore, since 2008, LULUCF once again becomes a key component for the 2nd commitment period. Developed countries stress that consensus must be reached on LULUCF rules before they can make any commitment to quantifying emission reduction after 2012.

(II) **Main issues and progress in negotiations on forestry**

Under the UNFCCC and the Kyoto Protocol, the main issues during a negotiation in forestry are REDD+ and LULUCF, with the former targeting forestry in developing countries and the latter on forestry in developed countries. Historically speaking, LULUCF had been under discussion several years ahead of REDD+.

1. Topics on LULUCF. Current negotiation on this topic is preceded on the basis of LULUCF rules during the 1st commitment period of the Kyoto Protocol. Key elements of LULUCF during this period are as follows: (1) If LULUCF activities in 1990 (base year) lead to carbon sources, then emissions from LULUCF activities and emissions from industry and energy in 1990 shall be put together as the basis (cardinal number) for determining the permitted emission quantity during the 1st commitment period for developed countries.[1] If LULUCF activities in 1990 (or base year) lead to carbon sink, then this carbon sink shall be allowed to be calculated in accordance with certain rules for offsetting the commitment to emission reduction. (2) Since 1990, carbon source/carbon sink from deforestation during 2008–2012 (1st commitment period) that are resulted from afforestation, reforestation, and anthropogenic activities must be included in the accounting. Only the net change during the 1st commitment period will be accounted, without deducting the net change of activities during the base year.[2] The result of accounting will be used for adding or reducing permitted emission. (3) Forest management, farmland management, pasture management, and vegetation recovery that have been organized by human beings since 1990 are subject to developed countries for inclusion in the accounting. Once decided, it can't be changed during the 1st commitment period. (4) If the net change accounting of carbon source/carbon sink from forest management during the 1st commitment period is carbon sink, then it should be offset by the following approaches: first, if afforestation, reforestation, and anthropogenic deforestation lead to a net change of carbon source during the 1st commitment period, the accounted carbon source from forest management can be used to offset those from afforestation, reforestation and anthropogenic deforestation, but each country may only use no more than 9 million tons of carbon to offset for this purpose each year; second, the remaining carbon sink from forest management after offsetting can be used for offsetting emissions from industry and energy sector, but there is a specific limit for each country to offset for this purpose. The limit is made up of two parts of carbon sink, namely forest management in developed countries and forest management project based on Joint Implementation. The specific limit for each developed countries is listed in the decision on accounting rules for LULUCF, namely the annex to Decision No. 16 of COP1 of the Kyoto Protocol. (5) Qualified LULUCF activities under the Clean Development Mechanism (CDM) during the 1st commitment period are only limited to afforestation and reforestation activities. The carbon sink from afforestation and reforestation projects for each developed countries under the CDM shall not exceed 1% of combined total emission from industry and energy sectors in the country concerned in the year of 1990.

At the request of developed countries, LULUCF was created during the negotiation of the 2nd commitment period under the Kyoto Protocol, with detailed discussion on how to revise LULUCF rules for the 1st commitment period. The main issues involved are as follows: whether to continue with the activity-based accounting methodology during the 1st commitment period or adopt an accounting methodology that is based on land use? How to revise the accounting rules that are in response to Article 3.3 of the Kyoto Protocol during the 1st commitment period, so as to increase

[1]The article is proposed by Australia.

[2]Generally known as gross-net.

the flexibility of logging and using of planting forest? Under Article 3.4 of the Kyoto Protocol, whether new activities will be included in the accounting during 2nd commitment period? Will these activities be mandatorily included in the accounting or can be included on a selective basis? How to calculate the carbon source/carbon sink caused by forest management during the 2nd commitment period? How to eliminate the impact of the interference of natural force majeure on the accounting outcome? Whether to set a limit for the amount of calculated carbon sink from LULUCF activities to be used for offsetting emission reduction commitment during the 2nd commitment period. Whether to add new and qualified LULUCF activities under the CDM, and whether to revise the definition of forest and afforestation or to add definitions of newly added activities?

After a series of negotiations, COP17 was convened in Durban at the end of 2011, in parallel with CMP7, during which parties reached the following consensus on LULUCF during the 2nd commitment period: continue with the activity-based accounting methodology during the 1st commitment period, but also agree to further discuss the accounting methodology that is based on land use; During the 2nd commitment period, calculation of carbon source/sink from LULUCF should follow the fundamental principle that is in line with that during 1st commitment period. The definitions that are related to forest, afforestation, reforestation, deforestation, vegetation recovery, forest management, farmland management and pasture management will remain unchanged; Agree to add "logging of wooden forestry product" and "wetland draining and wetland restoration" activities during the 2nd commitment period; Agree to include in the accounting all changes of carbon emission/carbon sink from afforestation, reforestation, deforestation, and forest management since 1990 in a mandatory manner, while changes of carbon emission/carbon sink from farmland management, pasture management, vegetation recovery, logging of wooden forest product, wetland draining, and wetland restoration are subject to individual countries to decide whether to have it included in the accounting; for activities that are included, countries should try to prove that they are indeed caused directly by anthropogenic factors since 1990. Under Article 3.3, it is also stated that forest planted from Jan. 1, 1960 to Dec. 31, 1989 can be logged. If plantation forest that is of the same area is made available on forest-free land on Dec. 31, 1989, and if it can be monitored and proved that the newly planted forest can reach the same carbon stocks as the logged forest during the log rotation period, then the forest logged from Jan. 1, 1960 to Dec. 31, 1989 can be accounted as part of forest management activity. Otherwise, it should be accounted as emission. Under Article 3.4, the method to account carbon emission/carbon sink of forest management during the 2nd commitment period is as follows: using the net change of carbon emission/carbon sink caused by forest management of related countries during the 2nd commitment period under the Kyoto Protocol to minus the reference level of carbon emission/carbon sink caused by forest management of related countries, thus obtaining the outcome; After that, using 3.5% of industrial and energy mission of related countries in 1990 as the limit to determine the actual increase and decrease of carbon sink or carbon emission of the emission permits of related countries. The method to account carbon emission/carbon sink from vegetation recovery, farmland management, pasture management, wetland draining, and wetland restoration is as follows: using the net change of carbon emission/carbon sink caused by vegetation recovery, farmland management, pasture management, wetland draining, and wetland restoration during the 2nd commitment period under the Kyoto Protocol to obtain outcome, which is then used directly to increase or decrease the emission permits of related countries. The method for accounting the logging of wooden forest products is "instantaneous oxidation" and "first-order decay functions". It is also subject to the relation between carbon emission of the accounted logging of wooden forest product and reference levels of forest management in various countries. When using "first-order decay functions" to calculate the carbon emission of the logged wooden forest products, it only covers domestic logging, domestic use, and paper exported from the country concerned to others, manmade plank and sawed products. Forest products imported from other countries are not included. When using "first-order decay functions" to calculate carbon emission of paper, manmade plank, and sawed products, the half-life is set at 2 years, 25 years and 35 years, respectively. It is allowed to deduct carbon emission of forest fire and pest damage caused by natural force majeure from the accounting result of carbon emission/carbon sink caused by forest management, but the specific location, year, interference type, and other information with respect to forest carbon emission caused by forest fire and pest damage resulted from natural force majeure must be provided. In addition, natural interference that is not caused by human being and is beyond the control of countries concerned must be proved. However, if rescue logging of forests interfered by forest fire and pest damage due to natural force majeure is necessary, and then the rescue logging is counted as emission. When accounting the carbon emission/carbon sink from afforestation, reforestation, deforestation, forest management, farmland management, pasture management, vegetation recovery, logging of wooden forest product, wetland draining and wetland restoration, the carbon stocks of surface life, belowground life, litter, dead and dying tree, soil organic carbon and logged harvest wood product shall be included. In addition to carbon pool of logged wooden forest products, for those that are not included in the accounting, countries must prove that they are emission sources; otherwise, they must be included in the accounting. When changing natural forest to plantation forest, all the carbon emission during the process must be included in the accounting. During the 2nd commitment period, only afforestation and reforestation activities can continue to be implemented as qualified CDM projects. The rules to be followed in the implementation are exactly the same as those during the 1st commitment period. With respect to whether other LULUCF activities should be counted as CDM projects, parties are still discussing this topic. Starting from 2012, in accordance with CMP7 decisions, parties began negotiations on the remaining issues of LULUCF: namely whether to adopt a comprehensive LULUCF accounting method, whether to include new LULUCF activities under CDM and set up corresponding implementation mode and process, how to resolve non-permanence risk of LULUCF projects under CDM, and how to use the additionality concept, etc. So far, no progress has been made in the negotiations on these issues.

2. Topics on REDD+. Since COP11 agreed to enlist "reducing emission from deforestation in developing countries: incentive principles" in the provisional agenda of the conference, starting from 2006, parties began technical discussion on science, social economy, technical methodology that are related to the topic under the scientific advisory body of the UNFCCC. The specific issues under discussion are as follows: the role of forest especially tropical forest in global carbon cycle, definition of forest, availability and quality of data, driving force and rate of deforestation, estimate of carbon stocks and forest coverage change and uncertainties, reducing non-permanence risk and leakage through bilateral and multilateral cooperation, promoting sustainable forest management and capacity building, funding mechanism, and relations between technology and incentive mechanism. Before COP13, the discussion is mainly on reducing emission from deforestation (RED). During COP13, parties had heated discussion over the scope of this topic. African countries stated that for the next step of discussion of incentive mechanism and policy measures, activity should not only target RED but also emission from the degradation of forest. China and India further noted that incentive mechanism and policy measure should also cover forest conservation, sustainable forest management, and carbon stocks increase since these activities can also contribute to mitigating global warming. Thus, they should be mobilized to ensure fairness and give full play of forestry of developing countries in mitigating global warming to the maximum. After protracted heated discussion, parties finally agreed that incentive mechanism and policy measure need to consider the role of forest conservation, sustainable forest management, and increase of carbon stocks. So far, RED is extended to cover activities including reducing emission from forest degradation, forest conservation, sustainable forest management, and increase of carbon stocks, and RED is also included in the "Bali Action Plan". After the COP14 at the end of 2008, REDD+ as an acronym is widely used.

Negotiation on REDD+ topics mainly covers the following: target, scope and guiding principle of REDD+, implementation measure, measureable, reportable and verifiable outcome, funding support of developed countries, forest governance and regime arrangement of developing countries, beneficiary of local community by REDD+ action, biodiversity conservation, promotion of effective local participation, equitable interest distribution, natural forest protection and prevention of carbon leakage, etc. After several rounds of negotiation, at COP18 at the end of 2012, parties reached general consensus on the scope and funding channel of REDD+, and requested "Green Climate Fund" to provide financial support for REDD+ activities. With the funding and technical support in place, developing countries may implement REDD+ activities by phases, including developing and implementing REDD+ national strategy, capacity building, pilot project, identifying reference level/reference emission level of REDD+ activities, setting up national forest monitoring system, gradually transition to full implementation of REDD+, and provision of outcome in accordance with measureable, reportable and verifiable requirement. They agreed to draw upon experience of current and future pilot activities and consider to set up a proper market mechanism to support REDD+ in developing countries. They agreed that in implementing REDD+ activities, they shall follow the principles of conserving biodiversity, equitable interest distribution, and protecting natural forest; they will also report and provide regular update on their compliance with these principles through national communication. They also agreed to set up reference level/reference emission level as the method to evaluate the effect of REDD+ activities in developing countries. The reference level/reference emission level shall follow IPCC guidance and need to be in line with carbon emission/carbon absorption of forest reflected in national communication. Developing countries are required to provide such information as creation of reference level/reference emission level to the Secretariat of UNFCCC, who will make it available at the website. Parties will provide regular update and receive technical evaluation. During COP19 at the end of 2013, parties adopted the Warsaw REDD+ framework. They reached consensus on the following issues: monitoring system, and measurement, report and verification guide of forest, driving force behind deforestation and forest degradation in developing countries, time and frequency of providing information on compliance with biodiversity protection, setting up technical evaluation guide on forest reference level/reference emission level, and supporting the funding for the "Warsaw REDD+ framework". This means that REDD+ activities have moved into the implementation phase.

(III) **Outlook of negotiation on forestry issues**

Looking back at the history of negotiation on forestry issues under the UNFCCC and the Kyoto Protocol, LULUCF targets developed countries, while REDD+ focuses on developing countries, but negotiations are mainly on technical issues like measuring, reporting, and verifying the carbon source/carbon sink change resulted from land use and land use change. Though certain consensus has been reached by parties on the technical issues involved in forest, this doesn't mean that the consensus identifies the best solution for measuring, reporting, and verifying the carbon source/carbon sink resulted from land use and land use change. Actually, current technology needs to be further improved.

Taking LULUCF rules as an example. LULUCF rules that are applicable during the 1st commitment period do have some shortcomings. Negotiations on LULUCF rules during the 2nd commitment period tries to solve some of these problems, but still far from reaching the expected target. The problems of current LULUCF rules are as follows: (1) According to current regulation, not all LULUCF activity and related carbon pool are required to be included in the accounting, which makes Annex 1 Contracting Parties of UNFCCC still enjoy some freedom in accounting activities or carbon pool that may cause carbon sink increase instead of accounting or calculating less the activities that lead to carbon emission. (2) Even for mandatory accounting of LULUCF activities, the calculation methodology for carbon source/carbon sink change is not unified. For example, the method for calculating carbon source/carbon sink change resulted from afforestation, reforestation, deforestation, and forest management, and developed countries also differ from the method for accounting carbon source/carbon sink change resulted from forest management. This makes accounting in different countries incomparable. (3) It is difficult to set up a continuous tracking system for LULUCF activities that are included in the accounting, and it is also very difficult to verify the result of carbon source/carbon sink change caused by accounting LULUCF active 1 cities. (4) The way to deal with carbon source/carbon sink change resulted from logging artificial forest doesn't comply with environmental integrity principle. (5) No sound solution is available so far with respect to leakage and non-permanence issues. (6) Further improvement is needed to identify natural interference and to eliminate from accounting result the carbon emission resulted from the process, etc. In reality, these problems reflect that forest, land use, and land use change cover wide areas. Countries differ a lot in terms of national status, forest condition, data, and capacity. Scientific study has yet to provide effective technical approach. With countries further improving data in land use and land use change activities and further enhancing technical capacity, especially with the further development of earth observation approach like satellite remote sensing, it is necessary to revise the way to estimate, report and verify carbon source/carbon sink change that is resulted from land use, and land use change through further negotiation.

The COP17 held at the end of 2011 in Durban is the turning point for negotiation. Parties agreed to start the negotiation through the Ad Hoc Working Group on the Durban Platform for Enhanced Action (hereinafter "Durban Platform"), which intends to reach an all-inclusive and legally binding agreement on combating climate change in 2015, and will be implemented starting from 2020. Though it is still uncertain about what elements should be included in the Durban Platform negotiation, parties have a general consensus that forest should play a key role in combating climate change. It can be confirmed that the forest will continue to play a critical part in tackling global climate change after 2020.

Therefore, it can be expected that forest-related issues will continue to be an integral part of climate negotiation, both at technical and policy level.

2 Main Policy Measures for China's Forestry to Cope with Climate Change

With the progress in the negotiation on forest-related issues, China continuously enhances its work in combating climate change in forestry. In 2003, the State Forestry and Grassland Administration (SFGA) set up the Office of Carbon Sink Management, which is under the Department of Plantation and Afforestation. In 2007, it changed its name to "Leadership Office of Combating Climate Change and Energy Conservation and Emission of SFGA". In 2008, the State Council clearly authorized SFGA to undertake "managing and increase and maintenance of carbon sink in forest ecosystem". In 2009, Premier Wen Jiabao stressed at the first taskforce meeting on forestry of the central government that, forestry has a special role to play in combating climate change. Developing forest must be regarded as a strategic choice for China to combat climate change. Therefore, China has mainly taken the following policies and measures:

First, the forestry is put in China's *National Plan on Climate Change* and voluntary mitigation activity before 2020. In June 2007, the State Council issued the *National Plan on Climate Change* in China (hereinafter "national plan"), requiring that forestry need to increase GHG for carbon absorption, forest ecosystem should be maintained and enlarged in its overall functions and a sound ecological environment should be created. In 2009, the Chinese government announced voluntary mitigation initiative by 2020, which included a requirement that by 2020, China's forest area and reserve should be increased to 40 million hectares and 1.3 billion cubic meters respectively on the basis of 2005.

Second, in the *12th Five-Year Plan for National Economic and Social Development* and the *Work Plan for Controlling GHG Emission during the 12th Five-year Plan*, there are binding indicators requiring that a new forest area of 12.5 million hectares should be added, forest coverage should reach 21.66%, and forest reserve should be increased to 600 million cubic meters. This will mount rigid and binding pressure on forest-related work for governments at all levels, thus proving assurance for achieving mitigation commitment made to the world.

Third, the *Action Plan for Combating Climate Change in Forestry* and the *Key Points in Combating Climate Change in Forestry during the 12th Five-year Plan Period* are released. The two documents call for combination of the following: forestry climate change and national voluntary GHG emission; forestry mitigation and adaptation to climate change; forestry area expansion, carbon reserve increase and forest quality improvement and carbon sink enhancement; forest carbon absorption and carbon emission control; government leadership, and social participation. The key tasks of implementing forestry mitigation, adaptation, and related capacity building were put forward, and a large amount of funding were made in afforestation, forest management, forest fire and pest control, and wetland protection and restoration.

Fourth, regions with favorable conditions are encouraged to launch CDM afforestation and reforestation projects. In light of domestic afforestation initiative and afforestation for carbon sink, regions are encouraged to explore a pilot project that incorporates forestry into low-carbon and carbon trading pilots. They are also encouraged to launch policy research on intellectual properties and trade in forest carbon. They should actively cooperate with related departments to promote national legislation process for combating climate change.

Fifth, actively explore the establishment of national forest carbon monitoring system and encourage related scientific research. In accordance with IPCC guides and the existing forest resource monitoring system in China, technical solution, technical guidance, and work plan should be set up with respect to national forest carbon monitoring system. The technical solution for national forest resource investigation should be improved, and investigation indicators should be enriched and adjusted to meet the data, methodology, and model requirement of setting up forest carbon monitoring system, measuring, reporting and verifying forest carbon status, and submitting national information notification.

Sixth, further enhance the capacity in forest governance of forest management department at all levels. Continue to protect and develop forest-related legislation, enhance enforcement, further improve the reform of forest right regime, and explore equitable carbon sink right and mechanism for related interest sharing; provide knowledge training and publicity campaign in climate change at grass-root level, so as to enhance the capacity of forest departments and people living around forest in participating and implementing climate change initiative; embracing the society by actively encouraging companies, groups, and individuals to participate in activities like plantation, afforestation, forest protection, protection and increase of forest carbon, and combating climate change.

Seventh, further improve the capacity of forest ecosystem in adapting to climate change. Combating climate change covers mitigation and adaptation. The key to adaptation is maintaining a sound forest ecosystem. More efforts should be made on researching the seed selection during afforestation, adjustment of forest stands structure, and ecosystem management against the background of climate change. Adaptation management should be incorporated in forest management planning and wild fauna and flora protection. Priority should be given to protecting vulnerable ecosystem, population, and species, improving forest fire and pest prevention and control system.

Eighth, actively and constructively participate in climate change negotiations and promote tangible bilateral and multilateral cooperation in forestry to combat climate change. Giving a bigger role to Asian-Pacific forest restoration and sustainable management network, and enhance exchange and cooperation in forest in Asian Pacific region. Actively encourage experts to go global and promote policy measures and best practice in China's forestry in Asian Pacific region and beyond.

3 Opportunities and Challenges Facing China's Forestry in Dealing with Climate Change

(I) Combating climate change brings new opportunity for China's forestry

First, against the background of climate change, the special role of forest in combating climate change has won wide attention, which will help the whole society to recognize the value of forest and the importance of forestry. This will further highlight forest's role and contribute to a better social atmosphere for forestry; second, the UNFCCC and the Kyoto Protocol give rise to global carbon market that offers new horizon for promoting forestry by innovative mechanism. The origin and development of carbon trading market have two implications: On one hand, price is set for carbon emission. Such pricing will be feedbacked to emitters, thus producing major impact on emission; on the other, the overall emission reduction cost of GHGs can be reduced. By participating in carbon trading, some capable companies may obtain certain benefit after achieving the emission reduction target. Forest carbon is also a part of global carbon trading, which means that carbon sink, with its ecological value, has already become tradable environmental products. For some reasons, the trade volume of global forest carbon is still limited, but this shows that it is quite possible to internalize the benefit of carbon sink externality through market mechanism. With the implementation of trade in forest carbon, it is also possible to achieve value compensation of eco-value for forest protection, biodiversity, water and soil conservation, water quality purification, and landscape offering. Such mechanism will contribute to combining the users and providers of forest eco benefit effectively. While providing benefits for companies sharing in forest eco-value, it also provides incentives for protectors and developers of forest resources. Third, with the deepening of international cooperation in combating climate change, China will shoulder more global obligations in this regard. Forestry has become an integral part for China to commit to mitigation initiative to the world, bringing new vitality to forestry. Meanwhile, giving full play to the forest's role in mitigation, China's development space is extended to certain extent, especially the international practice in carbon trading. Companies are encouraged to offset their emission by obtaining carbon sink through plantation and afforestation, which not only reduce their emission reduction cost but also make them undertake certain social responsibilities and raise their awareness to forest protection. This will contribute to expanding funding channels for forestry and promoting regime innovation of forestry. Fourth, giving full play of forestry in combating climate change not only involves afforestation, forest update, and management but also includes developing forest and wooden biomass energy to replace fossil fuel energy and chemical energy material. For example, using the fruit of forest producing oil energy to extract oil for biodiesel, using targeted nurturing of energy forest, litter of logging at forest area and scrap from wood processing for direct burning for power or heating, converting wood half cellulose to ethanol fuel, and using wood to replace certain fossil fuels in producing bricks, iron and steel, aluminum, glass and other raw materials. A UK study shows that using 1 ton of wood as building material will require 640 KWh power during production and transport and will emit 480 kg of CO_2. However, the energy required in producing and transporting 1 ton of bricks, cement, glass, steel, and aluminum is respectively 4, 5, 6, 24, and 126 times of that in producing and transporting wood materials of the same amount. Experts believe that mitigating global warming will mount the production cost of high-density energy products and will definitely increase the demand for wood and renewable resource to replace these products. This will not only contribute to emission reduction and provide new growth point for sustainable social and economic development, but also expand the development scope of forestry. Fifth, continue to participate in afforestation and reforestation projects under CDM. While introducing certain amount of funding for afforestation, it also provides experience in forest carbon measuring, monitoring, verification, and trade and enhances capacity of different regions in participating and implementing trade in carbon sink. It will serve as reference for further relying on market mechanism to promote eco-value compensation. This can be used to further encourage the whole society to participate in plantation and afforestation for carbon sink increase and combat climate change by voluntary contribution or other approaches. This will contribute to awareness-raising to forest protection and development and combating climate change, expanding funding channels for afforestation and greening initiatives for the time being, and further stepping up China's afforestation and greening.

(II) The impact of climate change also pose challenges to China's forestry

First, climate change will have a major impact on the productivity of forest, species distribution and ecosystem interference (fire and pest). Although such impact has both advantage and disadvantage, generally the latter surpasses the former. If the disadvantage can't be avoided, forest won't contribute to mitigating climate change, and will exacerbate the warming trend and further deteriorate forest growth. In recent years, global warming has led to exacerbated frequency and density of forest fire and pest damage, which makes it more difficult to protect and develop China's forest resource. Second, agriculture and forestry have close relationship. Climate change affects agriculture mainly in the fact that the overall food output in the future may be reduced. With the increasing population, this will mean that more forest or forest land are subject to be destroyed or be occupied for planting grain or developing husbandry, which will lead to competition for land use. In addition, industrial development and urbanization also require certain portion of land. This will not only mount pressure for managing forest and forest land, but also put on constraint to carbon sink increase through expanding forest area. Third, the impact of climate change on forest productivity, species distribution and ecosystem interference, if reflected in actual forestry work, mainly reflects in the fact that it affects seed selection and plantation practice in different regions. It will have a major impact on the provision of global wood, non-wooden forest products, and various forest services. Numerous climate and economic model research have shown that multiple benefits can be obtained in mitigating climate change through forest approaches. It can contribute to reducing the implementation cost of mitigating climate change, but will also bring about changes in global land use pattern. Against the background of climate change, how to balance the need between forest products and services and increase carbon sink to cope with climate change on one hand, and provide continuous, effective, and positive incentive for local operators on the other, requires forest departments to reform and innovate current policies, regimes, and systems. This is also a challenge for China's forest department. Fourth, according to the Food and Agriculture Organization of the United Nations (FAO), the amount of industrial log around the world totals 1.6 billion cubic meters each year (FAO 2005). By 2050, the data above may reach 1.9–3.1 billion cubic meters, with annual increase of 0.5–2.0%, and annual price increase of 0.5%. The intensity of logging may vary according to regions. With countries beginning to fully implement REDD+ initiative, the opportunity cost of logging and utilization will definitely increase.

Opportunity and challenge are the two sides of one coin. In principle, climate change will further promote the Chinese government to pay more attention to forestry and reform the current management regime and practice that are no longer applicable. This will bring in more driving force for China's forestry.

4 Potential and Trend Analysis of China's Forest Carbon

As the main body of terrestrial ecosystem, forest absorbs CO_2 in the atmosphere through metabolism for growth. CO_2 is consolidated inside forest or forest land in the form of biomass, which makes forest the main carbon pool on the earth. Forest carbon pool is made up of aboveground and belowground biomass, litter, dead and dying trees, and soil carbon pool. The biomass at aboveground and belowground, together with the total amount of litter, dead and dying trees, and carbon in soil carbon pool, made up forest carbon stocks. Such reserve may change subject to human being, natural reasons and time, the result of which may be either carbon sink or carbon source. When forest reserve shows increase during certain time, this means forest continues to absorb CO_2 and other GHGs in the atmosphere, namely forest becomes carbon sink. Otherwise it will be carbon source.

Since 1949, with China's political and economic development and change, China's forest also experienced changes from large-scale logging to large-scale afforestation. These changes definitely bring about transformation in China's forest carbon. Academician Jingyun Fang and others use the data from national forest resource survey of 1977–2003 for an analysis. The result is that, since the end of 1970s, the carbon pool of forest vegetation in China has increased significantly. National forest carbon pool has increased from 430 million tons of carbon in early 1980s (1977–1981) to 585 million tons of carbon in early 2000s (1999–2003), with annual carbon sink of 7.5 million tons of carbon. Forest carbon intensity per area is also up from 36.9 tons of carbon per hectare in early 1980s to 41 tons of carbon in 2003. In particular, during 1999–2003, the carbon sink of national forest reached 17 million tons of carbon per year, surpassing that of US forest vegetation, which stood at 11–15 million tons of carbon each year. Among these, China's plantation forest contributed over 80% to national forest carbon.[3] According to the calculation based on FAO's evaluation of global forest resource: the carbon reserve for China's forest in 2000 and 2005 is 5.0 GtC[4] and 6.1 GtC, respectively. The result of this general analysis is that the carbon stocks of China's forest vegetation, soil, and litter are 5–6, 20 GtC and around 0.8–0.9 GtC. Though the research result of the future trend in China's forest carbon may differ due to research approach, the general feature is shown as a downward trend before 1980s, and an upward trend after 1980s. From 1990 to 2050, the net carbon sink of China's forest will gradually increase. It is estimated that by 2050, the net carbon sink of China's forest accounts for 7% of GHGs emitted from industrial sector during the same time period.

Like forest distribution, China's forest carbon stocks are imbalanced across regions. The southwest and northeast enjoy the largest forest ecosystem carbon stocks, accounting for 24–31% of national total. The northwest and North China regions are relatively smaller in carbon stocks. However, since the implement of the protective forest engineering across northwest, central north and northeast at the end

[3]Estimate of Carbon Sink of Terrestrial Vegetation in China from 1981–2000, Fang Jingyun et al., Science China, 2007, Vol. 37, Issue 6.

[4]$1Gt = 10^9$ ton.

of 1970s, forest carbon stocks in these regions increased significantly. Carbon density of forest vegetation is the highest at alpine vegetation zone in Tibetan plateau, monsoon forest at tropical zone and needle/broad-leaved mixed forest at temperate zone, with carbon stocks per hectare reaching 200 tons. The evergreen broad-leaved forest vegetation at subtropical zone has the lowest carbon density at 120 tons of carbon per hectare. The northeast and southwest regions show relatively higher carbon stocks and carbon density during different survey periods. This has something to do with the fact that these regions mainly have northern forest with higher biomass stock and subalpine needle-leaved forest including spruce fir and larch. The east and central regions with artificial forest have relatively lower carbon density.

All in all, in the past 20 years, China's forest carbon has been increasing continuously. Forest is and will be a major carbon sink. However, ecosystem carbon sink/source has complicated process of change, which is affected by multiple factors and changes dramatically among years. Therefore, further exploration is needed with respect to forest carbon stocks and change of carbon sink/source in China's forest in the future.

5 Main Approaches to Reducing Climate Change Potential for China's Forestry

In mitigating climate change, the role of forestry is achieved through the following approaches: enhancing forest carbon sequestration, reducing forest carbon emission, increasing carbon storage in harvest wood products, and promoting carbon replacement in harvest wood products. Specific measures include enhancing forest carbon sequestration through plantation, afforestation, vegetation restoration, and sustainable forest management; reasonably controlling wood logging, reducing deforestation, preventing forest fire and pest, reducing forest emission; extending longevity of harvest wood products by increasing their utilization, extending carbon storage period of harvest wood products, and enlarging carbon storage capacity in harvest wood products; Using harvest wood products, especially forest logging and processed litters can be converted into energy to partially replace fossil fuel and using wood to replace fossil fuel materials. These all contribute to reducing fossil fuel consumption and carbon emission. China's forestry has certain potential in these aspects.

(I) **Enlarging forest area an increasing carbon sink through plantation and afforestation**

Compared with developed countries and some developing countries, China's forest coverage is relatively lower. China still has 57 million hectares of bald mountains and lands that are suitable for forestry, 54 million hectares of sand and desert that are suitable for forestry. A number of sloped farmland at 25° and uncultivated land can be used for plantation and afforestation. Meanwhile, there is still broader space for increasing the utilization rate of existing forest, developing forest network among farmlands, so as to enlarge forest coverage in China. According to the *Decision on Accelerating Forestry Development by the State Council*, the mid and long-term development goal is, by 2050, to increase China's forest coverage from 18.21% for now to above 26%. By then, the average carbon sink of forest in China will be greatly increased.

(II) **Increasing carbon sink by enhancing the quality of existing forest resource**

The average reserve of existing forest resource in China is 84 m^3/ha, and the average stands growth per hectare is 3.55 m^3, which is less than 1/4–1/3 of that in developed countries. Most forests are artificial and secondary type with lower biomass. According to expert analysis, the carbon stocks of existing forest vegetation resource in China only accounts for 44.3% of its potential reserve. By reasonably adjusting stands structure and enhancing forest management, it is quite possible to double the stands growth per area, thus greatly enhancing the carbon sink of existing forest vegetation.

(III) **Reducing forest carbon emission and enhancing forest capacity in mitigating climate change through enhanced forest protection**

By strictly controlling arbitrary appropriation and occupation of forest land that may lead to deforestation, forest carbon emission can be reduced. Historical inventory of China's forest resource has shown that there are over 1 million hectares of forest area lost each year in China due to arbitrary appropriation and occupation of forest land. Through scientific planning and strict enforcement, deforestation resulted from arbitrary appropriation and occupation of forest land can be curbed, with greater emission potential mobilized. During forest logging, low-intensity and low-interference logging should be adopted to protect forest vegetation and soil. This can reduce carbon emission that is caused by vegetation and forest soil that are destroyed by logging. By enhancing effective management over flammable objects in forest, setting up forest fire and pest early-warning system, and effectively controlling the frequency and implication of forest fire and pest, forest carbon emission can also be reduced.

(IV) **Improving forest capacity in mitigating climate change through wetland protection and preventing water and soil loss in forest**

Wetland soil is rich in abundant organic carbon. If destroyed, the stored organic carbon will be broken down and CO_2 and GHGs will be released to the atmosphere. China now has various wetlands (over 100 ha) totaling 38.48 million hectares, but a large number of these wetlands are deteriorated or occupied due to economic and social development. Enhancing wetland protection will reduce GHG emission resulted from wetland damage. Forest soil also stores a large amount of organic carbon, accounting for over 60% of carbon stocks in forest ecosystem. Research shows that organic substances in forest soil can oxidize and capture methane in the atmosphere, which is conducive to mitigating climate change. Enhancing bio measures and controlling forest land and soil loss will contribute to protecting forest soil, promoting and stepping up soil growth, converting nonforest soil into forest type and strengthening carbon sequestration capacity of forest soil.

(V) **Increasing the percentage of wooden biomass energy in replacement of fossil fuel and enhancing forest capacity in mitigating climate change through forest biomass energy**

By direct burning of biomass raw material, wood fiber can be hydrolyzed and converted into ethanol, pyrolysis, and gasification and utilizing the fruit of oil energy forest for producing biodiesel. They can partially replace fossil fuel and reduce GHGs emission. Research shows that 1000 tons of wood litters can be converted into 75–85 gallons of ethanol fuel or 550–650 KWh of electricity. According to an expert estimate, the existing forest resource in China can provide 300 million tons of biomass for industry energy raw material, which can replace nearly 200 million tons of standard coal. Meanwhile, the existing forest land, salty-alkaline land, reclaimed land from mines and other difficult-to-use land can be used specifically to foster energy-type forest, so as to increase the percentage of wooden biomass to replace fossil fuel. It will contribute to reducing China's GHGs emission.

(VI) **Extending wood longevity, increase carbon storage capacity of wood products through proper wood utilization will contribute to mitigating GHGs emission and climate change**

The energy used during wood production and process is significantly lower than GHGs emission generated during iron and aluminum production. Wood can be partially used to replace energy-intensive material, which not only can increase carbon storage but also reduce the higher GHGs emission during the raw material production of fossil fuel. According to international energy organizations, using wood structure to replace steel structure can cut the energy consumption from 300 to 100; using wood structure to replace reinforced concrete structure can cut the energy consumption from 800 to 100. A UK study shows that using 1 cubic meter of wood to replace the same amount of cement and brick can reduce 0.8 tons of CO_2 equivalent, which not only saves energy but also reduce pollution and GHGs emission. As long as wood products don't perish and burn out, they are critical carbon pool. According to Chinese experts, between 1961 and 2004, the carbon stocks of wood products in China totals 1.2–1.8 billion tons of CO_2 equivalent. Starting in 2000, the carbon stocks of wood products in China increase by 7 million tons per year. All of these will further enlarge forest's contribution to mitigating climate change in China.

5.1 Future of Forest Carbon Market

(I) **Carbon trading**

It is originated from the trade in pollution discharge permit of SO_2 in US, with its theoretical basis on the trading cost theory put forward by Ronald Coase of Chicago University in 1960s. In 1968, a Canadian economist John Dale developed the trade in pollution discharge permit on top of it. As a new environmental regime, pollution discharge trade was implemented first in acid rain control in US and achieved success.

In the negotiation under the Kyoto Protocol, US put forward carbon trading as a flexible mechanism for inclusion in the Kyoto Protocol. The essence of carbon trading is a policy tool for reducing emission cost. It is recognized by many because differences exist among various countries, regions, and companies in terms of emission reduction cost. Atmospherically speaking, the location of emission reduction doesn't matter. Reducing 1 ton of CO_2 in either China or Japan will contribute to the same effect on protecting the atmosphere, but the cost may differ. This provides a reasonable ground for carbon trading across companies, regions, and national borders, which will lead to slashing emission reduction cost at last. According to experts, CDM projects have helped developed countries to save US$113–169.5 billion in emission reduction cost so far. In particular, EU has cut down 70% of emission reduction cost by purchasing carbon emission permit from CDM projects.

Implementing carbon trading generally needs to consider the following conditions: First, government legislation, namely to set an upper limit to the generally allowed emission of GHGs in a country or certain regions, which is called "cap" in English. How to set this upper limit or quantity control requires science-based data analysis and it needs to be secured by legislation. Second, a sound allowance distribution system needs to be set up. It is quite interesting as to how to distribute the upper limit (the "cap") set by government to discharge companies, and it also needs to be implemented by legislation.[5] EU-ETS has exceeded the limit when issuing allowance for the first time. Third, a verification and evaluation system needs to be set up to verify whether companies exceeded their limit during discharge. Fourth, a penalty mechanism needs to be in place to clearly identify punishment measures for companies exceeding their limits.

Currently, international carbon trading covers allowance trade and project trade. Among these, allowance trade mainly includes assigned amount units (AAUs) set by the Kyoto Protocol and European Union allowances (EUAs) identified by EU-ETS. Project trade mainly covers emission reduction credit obtained from emission reduction and carbon sink projects. It mainly includes certified carbon emission reductions (CCER) generated by CDM projects and emission reduction units (ERUs) generated from joint implementation projects. The global carbon trading market can be divided into compliance and noncompliance types (or the regulated or nonregulated, Kyoto Protocol and non-Kyoto Protocol). In compliance market, there are Kyoto Protocol-based market and trade market based on national or regional agreement, such as EU-ETS. Noncompliance market is also known as voluntary market. It has no clear-cut background of government legislation. Companies, civil organizations and individuals, out of social responsibility and long-term strategic consideration, voluntarily participate in trading activities generated from GHS emission control. Compared to compliance market, its overall scale is relatively smaller.

The main advantage of carbon trading is that it can help to use economic means for mobilizing the initiatives of companies in emission reduction, setting price for carbon and urging companies to incorporate carbon emission into production cost.

[5]Currently, Ministry of Ecology and Environment has set up quantity control and distribution system for SO_2 and COD.

In particular, the object of carbon trading can be stock trade or futures trade. Financial sector is now viewing the potential opportunity and promoting various financial derivatives under the concept of carbon emission permit, which makes carbon trading very popular. However, whatever the situation is, observation and thinking are needed. Currently, many people around the world are opposing carbon trading. A number of famous experts who used to speak in favor of carbon trading now begin to doubt or reject carbon trading, saying that carbon trading can never bring any tangible emission reduction. It is not appropriate to use the trading mechanism for SO_2 emission permit in the field of trading in GHGs emission permits, since the former doesn't allow offset mechanism. SO_2 emitting companies are mainly power plant, which is relatively simple and easier to monitor their emission sources. SO_2 emission trade only covers SO_2 only, while carbon trading allows cross-region and cross-border trade. It covers 6 GHGs including CO_2, methane, N_2O, HFCs, CF_4, and SF_6. It doesn't have effective monitoring measure for emission source, and company's emission reduction quantity is generally estimated. It also allows using the emission reduction quantity generated from emission reduction projects that are implemented by regions and countries without emission limit for offset purpose, which makes carbon trading system more complicated than that of SO_2. The final result is quite uncertain.

Although dispute exists around the world regarding the effect of carbon trading on emission reduction, it is still a valuable market mechanism for encouraging companies and regions to participate in energy conservation, emission reduction and coping with climate change. In particular, under the current negotiation, developed countries try to propose carbon trading mechanism, which has the purpose of reducing their own emission reduction cost and shifting responsibilities of emission reduction, but the positive effect of carbon trading should be recognized.

(II) **Current status and future trend of forest carbon trading**

Forest carbon can offset the emission from industry and energy companies to a certain extent and become a low-cost choice for them. In order to secure competitiveness a build up a green image, many companies are willing to obtain carbon credits by plantation, afforestation, and forest protection, so as to help themselves with emission reduction. According to literature, the earliest forest carbon trading happened toward the end of 1980s. Before the Kyoto Protocol come into effect, trade activities that are related to forest carbon take the majority of carbon trading. These trades are mostly conducted directly by company themselves. These projects are implemented by consultant and NGO, thus the trading price is generally very low. When the Kyoto Protocol came into effect, the trade in carbon sink, which is based on the rules under the Kyoto Protocol and represented by CDM afforestation and reforestation, has won attention. However, according to the Secretariat of UNFCCC, by September 12, 2012, there are 4,626 CDM projects registered across the globe, with only 40 on afforestation and reforestation. The accounted emission reduction is less than 1% of that in CDM projects. In terms of voluntary market, although forest carbon trading accounts for a large share, the market itself is relatively smaller, thus the scale of forest carbon trading is also quite small in the global carbon market.

Although the scale of forest carbon trading market is small, it can't be denied that forest plays a major role in mitigating global warming. Without forest mitigation, it will be more difficult to achieve the target of no more than 2 degrees of global temperature rise. The market share of forest carbon trading is small, which is directly related to the intrinsically special feature of forest carbon and policy constraint thereby. First, forest carbon in the global market is project-based and is typical trade based on offset mechanism. Although market principles are generally the same as other types of projects, forest carbon projects encounter difficulties in setting benchmark, substantiating the additionality, resolving non-permanence, avoiding leakage, and measurement and monitoring. In addition, some are concerned that the low-cost emission reduction quantity generated from large amount of forest carbon projects may deal an impact on carbon market price, which is not conducive to mobilizing the initiatives of industry and energy sector in emission reduction and developing renewable energy. Thus, several policy limits are set for forest carbon-offsetting industry and energy emission. Under the Kyoto Protocol, a strict mode and process have been set for CDM afforestation and reforestation projects. These projects can be used to offset the emission each year and may not exceed 1% of source emission of base year. In particular, CDM afforestation and reforestation projects can only certify temporary carbon credit, which greatly discourage the interest of buyers. EU-ETS market completely refuses forest carbon trading. In New Zealand, domestic carbon trading market also refuses any trade in carbon credits generated from CDM afforestation and reforestation projects, but its forestry department is the first to join domestic carbon trading system in 2008. In addition, initial evaluation also reveals that after forestry coming aboard, the country is motivated in protection and afforestation, and the large amount of carbon credit generated can be sold to other developed countries. This greatly reduces the pressure on its industry and energy in terms of emission reduction. In addition, EU suffers economic downturn, US is slow to pass climate legislation. Climate change negotiation is protracted, especially under the Kyoto Protocol. Developed countries are weak in their emission reduction commitment during the 2nd commitment period. These all lead to reduced demand for carbon credit globally, and further cut down on the scale of forest carbon trading.

Before 2020, global forest carbon market won't be developing well. In particular, the compliance market based on the Kyoto Protocol will still be small in its trade volume. A key reason to this is that during the 2nd commitment period under the Kyoto Protocol, only EU clearly accepts the term. But EU has relatively lower level of emission reduction commitment for now, plus its economic downturn, there won't be much pressure in accomplishing the emission reduction commitment during the 2nd commitment period. Since the operation of EU-ETS, it has yet to accept forest carbon credit trade, and it is very unlikely that it will accept such trade during the second commitment period of the Kyoto Protocol. Since the effective demand for carbon credit is significantly insufficient globally, forest carbon credit will be difficult to increase in the market. In voluntary carbon market, the trade in forest carbon credit will continue to maintain growth momentum but will be controlled by market rules, thus the overall scale won't be large. With some countries implementing REDD+ initiative, part of carbon credits from these initiatives will move into forest carbon market. The fact that supply exceeds demand will continue to lead to downward price of forest carbon credit. With green climate fund established under the UNFCCC, it will provide support for REDD+ and will contribute to exploration of creating REDD+ market mechanism to some extent. It will further improve measurement and monitoring technology that are related to forest carbon. All of these will be conducive to promoting further growth of forest carbon trading. According to the *State and Trends of the Carbon Market Report 2011* issued by the World Bank, the number of REDD+ projects at national and subnational levels are increasing steadily. Before 2020, forest carbon will continue to play a certain role in re-regulated carbon trading market at national or regional level, and it will also take the majority in voluntary market.

After 2020, the development of global forest carbon market will depend upon the progress made in climate change negotiation. Judging from the current situation, forest is a key component of Paris Agreement, in particular, parties have agreed to further discuss a key issue, which is to set up a proper market for providing funding support for REDD+. Forest carbon trading at national or subnational levels (or project level) need to be considered in market mechanism. Global carbon market will be further encouraged by Paris Agreement, and forest carbon market will correspondingly be further promoted.

Although it takes quite a long time for forest carbon market to take shape, the value of forest as a critical asset has won the attention of private sector continuously. Using market mechanism to encourage private sectors to participate is an important policy tools for decision makers. In this process, with the market operating mode for forest carbon trading becoming more and more improved, forest carbon trading will play its role in the carbon trading system at both international and national levels.

(III) **Analysis of the future of domestic forest carbon trading market in China**

The Outline of 12th Five-Year Plan for National Economic and Social Development clearly states that more efforts will be made to actively explore the development of domestic carbon market. This key measure will draw upon market approach to achieve emission reduction targets, thus providing more choice for emitters. Currently, Beijing, Shanghai, Tianjin, Chongqing, Guangdong, Hubei, and Shenzhen are conducting pilot projects of trade in carbon emission permits. On the basis of that, domestic carbon trading system will be set up gradually, which will contribute to domestic forest carbon trading.

Before 2020, China's mission reduction is relevant emission reduction with voluntary feature. Under the current situation, it would be difficult for the government to release quantity control policies on CO_2 emission. Given that policy background, unless these provinces with pilot projects can issue provincial-level quantity control policies and allow forest carbon to participate in the trade system, it would still be difficult to promote the development of domestic forest carbon market. For the time being, the undergoing carbon trading in Shanghai doesn't specify whether to include forest carbon. Guangdong Province may incorporate forest carbon into its trade system, but it's not clear as to how this can be included. There is great uncertainty as to whether China will take absolute emission reduction initiative after 2020.

It is affirmative that forest carbon will definitely play its due role in undertaking the obligation of absolute emission reduction in the future, the share of which is subject to the rules coming out of international negotiations.

6 Related Work in Promoting China's Forestry to Cope with Climate Change and Forest Carbon Market

(I) Further raise the awareness of forestry management departments at all levels to combating climate change in this sector. Actively implement China's commitment to the world that by 2020, forest area and reserve will be increased to 40 million hectares and 1.3 billion cubic meters respectively on the basis of 2005. The Action Plan for Combating Climate Change in Forestry will be implemented, continue to promote afforestation, forest management, forest land and wetland protection, forest fire and pest control, and actively participate in international negotiations on climate change.

(II) Draft and set up a complete set of equitable and transparent forest carbon monitoring system at national and project level, including related technical guidance books, which should be in compliance with IPCC guides as much as possible. Continue to conduct scientific research and technical training, actively participate in the national implementation of tackling climate change.

(III) Actively participate in the design of domestic carbon trading regime, identify the specific measures to incorporate forest carbon in the trade and clarify the relation with industry and energy sector in terms of emission reduction, including the share that forest carbon can be used for offsetting source emission. Set up a management system that is all departments may participate and cooperate. Nurture the development of measurement, monitoring, certification, registration, accounting, and other intermediary and consultation activities that are related to the market, so as to gradually form up a set of sound market operating rules.

(IV) On the basis of carbon trading system in China, explore and set up a simplified operating procedure which encourages forest carbon to be included in provincial and national carbon trading system and try to reduce trade cost as much as possible. Enhance knowledge dissemination and capacity building that are related to forest carbon trading. Promote stakeholders that are involved in forest, especially community farmers, to effectively engage in forest carbon trading. Further communicate with and encourage companies, civil groups, organizations or individuals to voluntarily participate in forest carbon compensation initiatives by all means, so as to raise their awareness to climate change.

(V) While implementing afforestation, forest management, and forest carbon trading, attention should be paid to multiple benefits in protecting biodiversity. Explore the way to set up an equitable and sound mechanism for benefit sharing. Improve forest governance regime, including clear and stable forest intellectual property rights, and forest insurance regime. Reduce the risks related to forest management and forest carbon trading.

Reference

FAO (2005) Global forest resources assessment. Rome, IT

General Concepts and Development Process of Forest Carbon Projects

Caifu Tang and Jian Ma

Summary

The chapter briefs on the general concepts that are involved in forest carbon projects, such as afforestation/reforestation (A/R), project boundary, and forest carbon pool. It elaborates on the development process of CDM forest carbon projects and China's GHGs voluntary emission reduction project, respectively; compare cooperative development and independent development as two development modes. Using project design document (PDD) of CDM A/R projects as an example, it describes in detail 10 major steps in the development of forest carbon project, including project identification, project boundary determination, crediting period selection, baseline investigation, baseline scenario recognition and additionality demonstration, leakage, participatory rural appraisal, environment impact assessment, biodiversity survey, and A/R operation and design.

1 General Concepts in Forest Carbon Project

(1) Afforestation

Afforestation under CDM refers to the direct human-induced conversion of land that has not been forested for a period of at least 50 years to forested land through planting, seeding and/or the human-induced promotion of natural seed sources.

C. Tang (✉)
Sichuan Green Carbon Ltd., Chengdu, China
e-mail: 631592345@qq.com

J. Ma
Paradise Foundation, Beijing, China
e-mail: mj@pfi.org.cn

Z. Lu et al. (eds.), *Forest Carbon Practices and Low Carbon Development in China*, https://doi.org/10.1007/978-981-13-7364-0_3

(2) Reforestation

Reforestation refers to the direct human-induced conversion of non-forested land to forested land through planting, seeding and/or the human-induced promotion of natural seed sources, on land that was forested but that has been converted to non-forested land. For the first commitment period of the Kyoto Protocol, reforestation activities will be limited to reforestation occurring on those lands that did not contain forest on December 31, 1989.

(3) Forest

In CDM A/R projects in developing countries, the forest must be defined and meet the following criteria (UNFCCC 2003):

Minimum tree crown cover (or equivalent stocking level) 10–30%;
Minimum land area 0.05–1.0 ha;
Minimum height of tree 2–5 m.

China's criteria for forest under CDM:

Minimum forest tree crown cover 20%;
Minimum land area 0.067 ha;
Minimum height of tree 2 m.

(4) Project participants

Project participants refer to contracting parties and/or state-owned, collectively owned and/or private bodies authorized by contracting parties that are engaged in CDM A/R project.

(5) Designated operational entity (DOE)

Bodies that are accreditated by CDM Executive Board (CDM-EB), appointed by COP/MOP, and are authorized to review proposed CDM projects and to verify and certify the removals of anthropogenic GHGs emission reduction. Unless specially specified by DOE Executive Board, a DOE can't undertake the verification and certification of the same CDM A/R projects.

(6) Project boundary

Project boundary geographically delineates the afforestation or reforestation project activity under the control of the project participants. The project activity may contain more than one discrete area of land, but each area of land should have a specific geographic boundary. This boundary doesn't include land between two and among different land area.

(7) Forest carbon pool

Forest carbon pool covers above-ground, below-ground biomass, litter, deadwood, and soil organic carbon. IPCC Good Practice Guidance provides a clear definition for five major carbon pools:

Above-ground biomass: All living biomass above the soil including stem, stump, branches, bark, seeds, and foliage is expressed in oven dry weight. In cases where forest understorey is a relatively small component of the above-ground biomass carbon pool, then it can be neglected, but consistency must be kept with respect to calculation and monitoring.

Below-ground biomass: All living biomass of live roots. Fine roots of less than (suggested) 2 mm diameter are often excluded because these often cannot be distinguished empirically from soil organic matter or litter.

Litter: All nonliving biomass with a diameter less than a minimum diameter chosen by the country (for example 10 cm), lying dead, in various states of decomposition above the mineral or organic soil. This includes the litter, fumic, and humic layers. Live fine roots (of less than the suggested diameter limit for below-ground biomass) are included in a litter where they cannot be distinguished from it empirically.

Deadwood: All nonliving woody biomass not contained in the litter, either standing, lying on the ground, or in the soil. Deadwood includes wood lying on the surface, dead roots, and stumps larger than or equal to 10 cm in diameter.

Soil organic carbon: Organic carbon in mineral and organic soils (including peat) to a specified depth is chosen by the country and applied consistently through the time series. Live fine roots (of less than the suggested diameter limit for below-ground biomass) are included with soil organic matter where they cannot be distinguished from it empirically.

(8) Baseline net greenhouse gas removals by sinks

Baseline net greenhouse gas removals by sinks refer to the sum of the changes in carbon stocks in the carbon pools within the project boundary that would have occurred in the absence of the afforestation or reforestation project activity under the clean development mechanism (CDM).

(9) Actual net greenhouse gas removals by sinks

Actual net greenhouse gas removals by sinks refers to the sum of the verifiable changes in carbon stocks in the carbon pools within the project boundary, minus the increase in emissions of the greenhouse gases measured in CO_2 equivalents by the sources that are increased as a result of the implementation of the afforestation or reforestation project activity, while avoiding double counting, within the project boundary, attributable to the afforestation or reforestation project activity under the CDM.

(10) Leakage

Leakage refers to the increase in greenhouse gas emissions by sources, which occurs outside the boundary of an afforestation or reforestation project activity under the CDM, which is measurable and attributable to the afforestation or reforestation project activity.

(11) Net anthropogenic greenhouse gas removals by sinks

Net anthropogenic greenhouse gas removals by sinks are the actual net greenhouse gas removals by sinks minus the baseline net greenhouse gas removals by sinks minus leakage.

(12) Baseline scenario

Baseline scenario refers to the scenario that reasonably represents the sum of the changes in carbon stocks in the carbon pools within the project boundary that would have occurred in the absence of the proposed project activity.

(13) Additionality

A CDM project activity is additional if anthropogenic emissions of greenhouse gases by sources are reduced below those that would have occurred in the absence of the registered CDM project activity.

(14) Certified carbon emission reduction (CER)

CER represents a reduction of 1 ton of CO_2 equivalent that is certified and issued.

(15) Non-permanence

Non-permanence refers to the scenario that the carbon reserve from CDM A/R project can be reversed due to natural or anthropogenic reasons.

(16) Temporary CER (tCER)

Temporary CER refers to a CER issued for an afforestation or reforestation project activity under the CDM, which subject to the provisions of section K below, expires at the end of the commitment period following the one during which it was issued.

(17) Long-term CER (lCER)

Long-term CER refers to a CER issued for an afforestation or reforestation project activity under the CDM, which subject to the provisions in section K below, expires at the end of the crediting period of the afforestation or reforestation project activity under the CDM for which it was issued.

(18) Crediting period

Crediting period refers to the period during which net anthropogenic greenhouse gas removals by sinks generated from CDM A/R projects that are certified or verified by designated operational entities for the purpose of lCER or tCER. The crediting period shall not exceed the operating period of the CDM A/R project.

(19) Stakeholder

Stakeholder refers to the public, including individuals, groups, or communities affected, or likely to be affected, by the proposed clean development mechanism project activity.

(20) Verification

Verification refers to the periodic independent review and ex post determination by the DOE of the net anthropogenic greenhouse gas removals by sinks achieved, since the start of the project, by an afforestation or reforestation project activity under the CDM.

(21) Certification

Certification refers to the written assurance by a DOE that an afforestation or reforestation project activity under the CDM achieved the net anthropogenic greenhouse gas removals by sinks since the start of the project, as verified.

(22) Registration

Registration refers to the formal acceptance by the Executive Board of a validated project as an afforestation or reforestation project activity under the CDM. Registration is the prerequisite for the verification, certification and issuance of tCERs or lCERs relating to that project activity.

2 Development Process of Forest Carbon Project

2.1 Operating Period of CDM A/R Project

The operation of CDM A/R project covers project development and project implementation. The operating period of CDM A/R project includes seven steps (Fig. 1).

1. **Project design**

Project participants organize project design and complete PDD.

(1) Draft development plan

According to the proposed project target and initial project scale, project participants draft project development plan, identify development approach, organizational structure, and time arrangement. They can also, based on their capacity and project risk, identify project partners, including CER buyers, project development intermediaries (such as consulting company), and lawyers. Project developers can be directly contracted by project owners or CER buyers, and development cost can either be covered by the owner or be borne by related parties (such as consulting company, carbon buyers, etc.). CDM lawyer can be recruited in project development to effectively reduce project risk.

(2) Initial selection of project location

Project owner and project developer, in accordance with the conditions set out in CDM A/R, select initial project location and organize project development experts

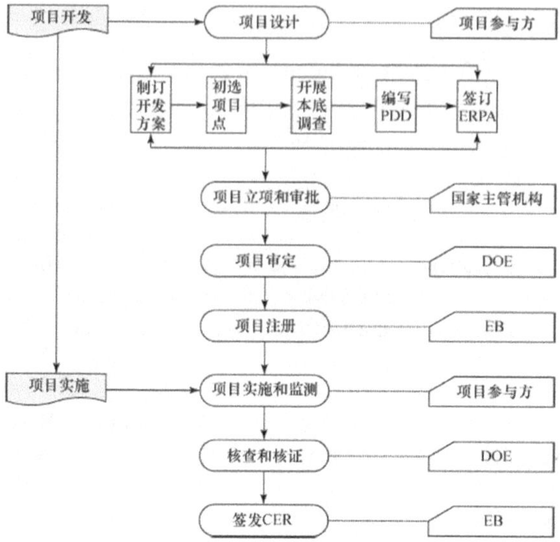

项目开发	Project development
项目设计	Project design
项目参与方	Project participants
制定开发方案	Draft development plan
初选项目点	Initial selection of project location
开展本底调查	Conduct baseline investigation
编写 PDD	Draft PDD
签订 ERPA	Sign ERPA
项目立项和审批	Project review and approval
国家主管机构	National authority
项目审定	Project validation
项目注册	Project registration
项目实施	Project implementation
项目实施和监测	Project implementation and monitoring
核查和核证	Verification and certification
签发 CER	CER issuance

Fig. 1 Operating period of CDM A/R project

and buyers to conduct tour studies and analyze the location selected: initial assessment of whether the project meets CDM A/R project requirement; identify whether a CDM A/R methodology is available in relation to the project type; comprehensively evaluate the cost–benefit, assess the feasibility and necessity of developing the project into a CDM A/R project, draft PIN, and submit CDM Project Notification Form to NDRC for record purpose.

(3) Conduct baseline investigation

Project developers, in accordance with selected methodology and CDM A/R modalities and procedures, draft technical methods for baseline investigation, participatory rural appraisal and biodiversity investigation, set up investigation team and conduct technical training, organize investigation staff to collect basic information of the project area, carry out investigation of the boundary, carbon baseline, biodiversity, and environmental impact on the project area. Initiate a participatory assessment of the community involved in the project, and obtain all information and documents required for drafting project design.

(4) Draft PDD

If the proposed project already has an applicable methodology, then project developer will, in accordance with the methodological requirement, draft PDD following the format set out by CDM-EB. PDD mainly includes project overview, project baseline selection, demonstration of project additionality, quantitatively estimate of project emission reduction, project environmental impact, stakeholder assessment, and project monitoring plan. During PDD drafting, project developer is responsible for providing all the supporting documents to cooperate with project development intermediaries in document collection and survey.

(5) Sign Emission Reduction Purchase Agreement

Emission Reduction Purchase Agreement (ERPA) is the legal basis of CDM projects for identifying the rights and obligations of CER buyer and seller. It is also the core of the CDM project. If a project is developed as a bilateral type, buyer and seller must reach consensus on the main terms (such as purchase amount, purchase price, delivery time, delivery method, project participants, and applicable law and resolution of dispute) of ERPA before submitting the project to NDRC for approval.

2. **Project review and approval at the host country**

NDRC is the national authority in China. CDM A/R projects need to be officially approved by NDRC. Upon completion of project design, project owners first need to file applications to national or provincial forestry management authorities where they are located for project review; after obtaining the approval, they will further apply for CDM projects from provincial development and reform committee where they are located. Certain state-owned companies can directly apply for CDM projects from NDRC if they are accredited by NDRC to do so.

The provincial development and reform committee where projects are located will submit all project application documents, together with initial comments to

NDRC within 20 working days, and they can't reject the application filed by project implementation bodies under any excuses. For applications that have incomplete documents or don't conform to the required format, the committees concerned shall request the applicants to supplement on-site or notify the applicants within 5 days of what is further needed.

NDRC takes in applications and organizes experts to review the case. The review duration shall not exceed 30 days. After the review, NDRC submits to project review board for further certification. NDRC will, based on the comments of the board, issue a letter of approval together with the Ministry of Science and Technology (MOST) and the Ministry of Foreign Affairs (MOFA). For projects that have been approved by the project review board, the approval procedure shall be started within 20 working days (excluding review time); for projects that have been approved by the board but need improvement and revision, the approval procedure shall be started after receiving the revised and improved documents from project implementation bodies, in cooperation with MOST and MOFA; for projects that have been refuted by the board, no approval procedure shall be started.

3. **Project validation**

CDM A/R projects are subject to the review of designated operational entities with respect to project activities (including project designs). After obtaining the letter of approval from government or in the meantime, project developers need to identify designated operational entities that have been authorized by the Executive Board or have the qualification of verifying and certifying A/R projects to sign an agreement. The latter is also responsible for reviewing and issuing a review report, upon completing these procedures; the project will become a legalized CDM project.

After receiving project review application, the designated operational entities will publicize PDD on the website of the Executive Board for 30–45 days; then it will conduct an independent assessment of the proposed CDM A/R project according to PDD, which usually take 2–5 months. After passing the review, the designated operational entity will issue its comments and submit the project to the Executive Board for approval and registration.

4. **Project registration**

After the approval of NDRC and qualified by a designated operational entity, a project will be submitted by the designated operational entity to the Executive Board for approval and registration. The Executive Board will publicize project information at its website for 8 weeks upon receiving registration application. Unless one contracting party of the project or at least three members of the Board raise the request to reexamine the CDM project concerned, the project will automatically pass the registration. If the project is rejected by the CDM-EB, applying company may revise the document and resend the application.

5. **Project implementation and monitoring**

Project participants implement project activities and carry out monitoring over the implementation status based on the monitoring plan proposed in PDD. They also collect related data, draft documents and submit the monitoring report to the designated operational entity that is responsible for verifying and certifying the quantity of project emission reduction.

6. **Verification and certification**

The designated operational entity verifies the CER generated from project activities and issue certification report. While verifying and certifying the quantity of project emission reduction, regular CDM projects are not allowed to invite operational entities that validated the project.

7. **CER issuance**

The designated operational entity submits the certification report to CDM-EB to request the CDM-EB to issue the same amount of CER as the verified quantity of emission reduction. Upon receiving the request for CER issuance, if there is no case of three and more members of the Executive Board oppose or any one government of the countries concerned objects, then the same amount of CER as the verified the quantity of emission reduction shall be granted within 15 days. After deducting 2% of CER, the remaining will be transferred to designated "account" according to the distribution plan set out in the agreement of project participants.

2.2 Operating Period of Voluntary Emission Reduction Project

According to the *Interim Measures for the Administration of Voluntary Greenhouse Gas Emission Reduction Transactions* (No. 1668 [2012] of the NDRC), the operating period of voluntary forest carbon project includes the following six steps (Fig. 2).

Project design: Project participants organize project design, complete PDD, and project owner submits the document for review by the reviewing bodies certified by NDRC.

Project review: The review bodies certified by NDRC are responsible for reviewing the trade in voluntary emission reduction, and issue review report.

Project registration: Project owner apply for registration of voluntary emission reduction at NDRC or development and reform committee at provincial autonomous and municipal levels. NDRC contracts experts to complete technical assessment within 30 working days. Other authorities will finish reviewing the registration process within 30 working days based on expert's comments.

Project implementation and monitoring: Project owner implements project activities according to project design, and works by itself or contracts a consulting company to draft monitoring report.

Certification of the quantity of emission reduction: Project owner request certification bodies under NDRC to certify the generated the quantity of emission reduction, and issue certification report for the quantity of emission reduction.

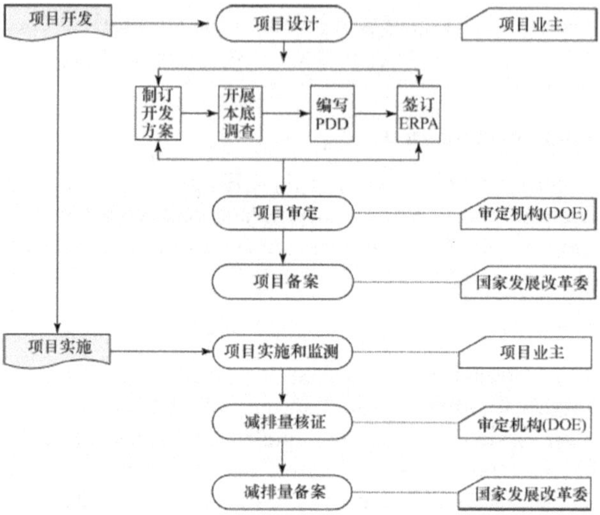

项目开发	Project development
项目设计	Project design
项目业主	Project owner
制定开发方案	Drafting development plan
开展本底调查	Conducting baseline survey
编写 PDD	Drafting PDD
签订 ERPA	Signing ERPA
项目审定	Project review
审定机构（DOE）	designated operational entities
项目备案	Project registration
国家发展改革委	NDRC
项目实施	Project implementation
项目实施和监测	Project implementation and monitoring
项目业主	Project owner
减排量核证	Certification of the quantity of emission reduction
减排量备案	Registration of the quantity of emission reduction

Fig. 2 Operating period of voluntary emission reduction project

Registration of the quantity of emission reduction: Based on the certification report, the project owner applies for the registration of the emission reduction quantity from NDRC. NDRC contracts experts to complete technical assessment within 30 working days, and finish reviewing the registration process within 30 working days according to expert's comments. The recorded quantity of emission reduction is called CCER, which can be traded at the trading institutions under a registration scheme.

3 Development Steps of Forest Carbon Project

Judging from the existing practice of forest carbon projects, the development modes generally covers independent mode and cooperative development mode (Table 1).

The main task of developing forest carbon project is to complete PDD, plus related supporting materials in the meantime. PDD must be prepared strictly according to the project design guide and the latest version of methodology. A CDM A/R project design is used hereby as an example to introduce the main steps in forest carbon project development.

3.1 Select Project Location

The implementation site of A/R projects should prioritize areas with the critical ecological region and vulnerable eco-environment. Following the requirement that CDM A/R project must be on eligible land, a project shall have clear rights and

Table 1 Development modes of forest carbon project

Development mode	Developer	Project type	CER sales mode
Independent development	Project owner	Unilateral project	Market sale
	CER buyer	Bilateral project	Signing purchase agreement
	Project intermediary	Unilateral or multilateral project	Market sale or signing purchase agreement
Cooperative development	Project owner and CER buyer	Bilateral project	Signing purchase agreement
	Project owner and project intermediary	Unilateral or multilateral project	Market sale or signing purchase agreement
	CER buyer and intermediary	Bilateral project	Signing purchase agreement

ownership, be suitable for tree growth and contributes to biodiversity, preventing desertification and local socioeconomic development.

Project participants must use transparent information to prove that the site selected for the project is a no-forest land since December 31, 1989. If organic carbon pool in the soil is to be dismissed conservatively, project participants need to substantiate the degraded feature of the site according to the "Tool for the identification of degraded or degrading lands for consideration in implementing CDM A/R project activities" formulated by the CDM-EB.

3.2 Identify Project Boundary

Project participants can use satellite images from different time periods to select the initially selected site that is consistent with project standard, and assess its eligibility through forest map, forest distribution map, land use, or land cover map from different time periods; meanwhile, on-site field tour is needed, GPS is to be used to record coordinates of inflection points along the boundary of the site and map out slopes on the landscape chart, GIS software is be used to generate land distribution chart. In addition, forest right certificate or land use agreement is needed as a proof that the project owner possesses all land rights or land use rights of land parcels with the boundary of CDM A/R project.

3.3 Select Crediting Period

The crediting period of CDM A/R project starts from the date when CDM A/R project activities commence. The crediting period has two choices:

Renewable crediting period: Maximum 20 years, renewable twice at most. The condition for renewing is that, for each renew, DOE shall determine and notify the Executive Board that the original baseline scenario is still effective, or that it has been updated in accordance with applicable new data.

Fixed crediting period: Maximum 30 years, non-renewable.

3.4 Baseline Investigation

1. **Select carbon pool**

CDM A/R project cover such carbon pools as above-ground biomass, below-ground biomass, litter, dead and dying trees, and soil organic substances.

When selecting a carbon pool, the first thing is to consider whether the net emission source is available; the second is to consider cost-effectiveness, uncertainty, and conservativeness of monitoring. If it can be proved from transparent and verifiable information that one or several carbon pools are not net emission source, then project participants may choose not to measure and monitor these carbon pools. Since great uncertainties exist in carbon stock change of afforestation/reforestation, the selection of carbon pool, the estimates, and monitoring of carbon stock change must be managed in a conservative approach. As soon as one carbon pool is selected, it must be measured and monitored continuously during the crediting period.

2. **Baseline stratification**

In order to improve the estimate, measurement and monitoring accuracy of carbon stock change in the baseline scenario and project scenario, and to reduce uncertainty, the selected CDM A/R projects need to be stratified.

The process of stratification is not subject to the size and spatial distribution of the project. The large size of land or several separate small lands can be viewed as one stratum. Baseline stratification is generally based on land use and vegetation of the project concerned before the project starts, and mainly consider the following factors:

Land use approach (such as fallow land, barren land).

Scattered trees and density (tree per area), prioritized tree species, and ages.

Type, height, and coverage of shrub vegetation (generally divided into a large shrub and small shrub, or dense distribution, scattered distribution).

The first thing to be considered in stratification is to identify whether scattered trees are available, followed by shrub then herbaceous vegetation.

3. **Sampling design**

In principle, there should be 1 sample plot per 50 ha. At least 10 sample plots should be randomly picked out from each baseline carbon strata for examination, with the examination criteria of ±20% accuracy at 90% reliability level. If the above accuracy rate is not reached, then temporary sample plots need to be added.

The central point of the sampling plot is determined by a random method, using Arcview's Random points module to randomly set central points for sample plots.

4. **Baseline measurement**

Herbaceous: herbaceous plant is not subject to baseline biomass survey.

Shrubs: including small mixed bamboo field with a diameter at breast height (DBH) less than 2 cm. Shrub coverage and average height should be surveyed and kept on record.

Scattered trees: measuring the crown cover of scattered trees on the sample plot, as well as the average height and DBH for each species with DBH over 5 cm.

5. **Natural regeneration survey**

Where naturally regeneration occurred, natural regeneration survey needs to be conducted. According to the *Guide on Measuring Land-based Carbon* developed by Winrock International, the trees on regenerated sample plots can be divided into three classes: seedling: height less than 30 cm; junior trees: height from 30 cm to 1.3 m; small trees: height over 1.3 m, DBH less than 5 cm.

6. **Land status survey**

Field survey is conducted on the selected project lands. Survey factors include terrain, landscape, water and land loss, soil, vegetation coverage, land ownership, land use history and current status, and main disaster and threats.

3.5 Baseline Scenario Identification and Additionality Demonstration

Conventional CDM A/R projects need to use the "Combined tool to identify the baseline scenario and demonstrate additionality in A/R CDM project activities" approved by CDM-EB to identify baseline scenario and prove the additionality of CDM A/R project.

When setting a baseline scenario, national or departmental policies and regulations that are conducive to afforestation/reforestation since the methodology and procedure adopted on November 11, 2001 under the Marrakech Agreement won't be considered, namely baseline scenario may presume there are no such policies and regulations. With respect to a non-proposed CDM A/R project, if there has been a certain scale of afforestation/reforestation projects in the past several years, and they are irrelevant with the policies in the wake of the cut-off date of November 11, 2001, plus that current or future social economy, environment, and policy are not expected to change significantly, then the afforestation/reforestation projects need to be considered as a baseline scenario. Generally, it is needed to assess the afforestation/reforestation rate in the baseline scenario, and substantiate that the afforestation/reforestation rate won't accelerate with any proposed CDM A/R projects.

When substantiating baseline scenario and additionality, it is needed to consider the natural regeneration. If natural regeneration is present, then the baseline scenario should be counted as a baseline scenario.

Additionality assessment needs to substantiate the barriers for proposed CDM A/R projects in terms of investment, funding, technology, and policy.

3.6 Leakage Investigation

CDM A/R methodologies only consider leakage of agricultural activities. Agricultural activities include anthropogenic cultivation and grazing. Cultivation refers to working the land with the purpose of crop production (such as grain, food, feed, fiber, and oil crop), including collecting and harvesting of crops. Grazing refers to land management system with the purpose of husbandry. If the land used to be farmland or husbandry land before the project starts, and remains so in the baseline scenario, then the proposed CDM A/R projects may cause a reduced output of crops or husbandry. This will lead to baseline cultivation or husbandry activities to be shifted beyond project boundary, resulting in increased GHGs emission outside project boundary (such as deforestation for cultivation land), hence the leakage of agricultural activities. When investigating into leakage, priority should be to given to identify whether the increase of GHGs emission caused by shifted agricultural activities is prominent, if not, leakage will be 0; otherwise, the methodology will be adopted to make estimate and monitoring.

3.7 Participatory Rural Appraisal

Participatory rural appraisal (PRA) is a method used to analyze rural issues and work with local stakeholders in seeking a solution. It utilizes broad and visible solution for collective analysis, with an eye to solving social and environmental issues from spatial and time dimension. As an effective approach to collecting stakeholders' opinions, PRA adopts participatory decision-making and planning mechanism to empower stakeholders with rights to information, participation, decision-making, and supervision, so as to incorporate them in project decision, design, implementation, and supervision and benefit from the process, especially the participation of such vulnerable groups as female and poor population. Information collected by PRA contributes to analyzing the impact of the proposed projects on social economy and environment, and helps to identify alternative land use solution and substantiate baseline scenario and additionality.

1. **Survey object**

The object of PRA includes project-covered villages, communities, and typical farmers (including village cadres, afforestation leaders, female representatives, poor households, rural pace-setters in cultivation), neighboring natural reserves, forest parks, governments, and NGO.

2. **Survey content**

The main content of survey covers the following: general information of community and villagers, socioeconomic activity, history, and current status of society and economy, distribution and utilization of natural resources (mainly land resource), social, economic, and environmental issues facing stakeholders, farmers' expectation and requirement on land use in the future and vegetation recovery project, as well as public awareness to environment.

3. **Survey step and approach**

(1) Team setup

Team members are made up of three–five professionals with various backgrounds (forestry, sociology, ecology, etc.) and familiar with PRA, so as to improve work efficiency and quality in designing the survey.

(2) Training organization

All design staff is required to attend technical training, with the purpose of identifying task, unifying methodology, and implementing responsibilities.

(3) Survey preparation

Prepare and print out community campaign materials, questionnaire, and other survey tools, such as terrain map, camera or video recorder, roll up tape or measuring rope, metal roll up tape, GPS, calculator, and drawing pen.

Collect available data and information on project county (city, reserved area) and town (township, site), especially chart, statistics form. Obtain general information, historical documents, and future development plans on natural geographical condition, socioeconomic condition, land use status. and forest resources.

Identify the objects of survey and towns (township, site) and villages (natural village) selected for the random survey, draft work plan, confirm team setup and responsibility, and arrange a schedule.

Contact counties (city, reserved area) concerned to confirm the towns, townships, villages, reserved areas, government departments (forestry bureau), and local NGOs to be surveyed, get clear with time, content, and team organization.

(4) Survey assessment

Field study: Field study is conducted together with villagers and village representatives to gain direct access to information on the general condition of project land and community resource use, which is useful for analysis and project planning. The main content covers the following: study and survey land ownership, land use, soil, vegetation, water and land loss, grazing, and collection; map out community resource map, identify project location and boundary; discuss future intention of use, barriers (funding, technology, and market) facing the expected land use and afforestation species suitable for the project land.

Semi-structured interview: Based on the background information already obtained, the survey team will draft an interview outline in connection with project objective and content. Key persons (village cadres, local elites, respected seniors, officials of companies, governments, and NGOs, head of minority groups), village representatives (10% from each village) and key groups (by gender, age and economic activity) will be interviewed. The interview will cover the following issues: land use history (policy change of forest land and timber, forest resource use in major historic phases); forest land and wood resource (scale, distribution, operation, use, effect, etc.); forestry rights and ownership (ownership, right to use, disposal and distribution, and record of issuing ownership certificate); forestry resource management (management and maintenance regime, rural contract and agreement, collective forestry reform, and benefit distribution); condition of suitable forest land (scale, distribution, formation, change trend, status of use, rights, and ownership); willingness of afforestation (mode of plantation, investment trend, selection of seeds, difficulties, and issues) and community livelihood (current status and issues, rural cooperative and economic organizations within communities, and their major events and outcomes).

Community meeting: Community members are organized for a meeting in light of survey content and the intended project planning. The scale can be either big or small. It can be organized by village or villager group, with everyone participated and at least one from each household should be present; it can also organize village representatives to participate in community meeting with respect to actual condition, but consideration should be given to gender, ethical groups, cultures, financial status, plus related stakeholders (rural government and rural forestry bureau should be present). Project background, objective, content, and mode of participation should be briefed to villagers during such meeting. Effort should also be made to facilitate sufficient discussion among villagers on land use expectation, barriers, mode or participation, seed selection, implementation, input, management, and maintenance, so as to reach consensus.

Questionnaire: Using a questionnaire to solicit information and comment from those being surveyed. The questionnaire can be used together with villager or village representative meeting, semi-structured interview, and other modes of participation.

(5) Summarization and analysis

Compiling survey and assessment documents, including written record and video record, so as to make it complete and in full, with clear and credible recording.

3.8 Environmental Impact Assessment

According to CDM A/R project modalities and procedures, project participants must submit to EB-approved DOEs environmental impact assessment reports including influence on biodiversity and the nature and ecosystem. If the managing body of the country where projects locate believe that it poses a potentially negative impact on the country's sustainable development goals, project participants need to conduct environmental impact assessment according to that country's required procedure. The description of the abovementioned regulations must be reflected in PDD.

According to the *Management Catalogue of Environmental Impact Assessment for Construction Projects* by Ministry of Ecology and Environment (effective as of October 1, 2008), managing bodies of environmental protection carry out categorized management over forest projects by categories of economic forest base, forest logging and desert prevention and control engineering. Namely, with respect to economic forest base, if a project falls under "raw material forest base", constructors shall submit to environmental protection bureau an environmental impact report compiled by an agency engaged in environmental impact assessment; if it is other types of project instead of "raw material forest base", it needs to submit an environmental impact form. With respect to forest logging project, if a project falls under "clear cutting", constructors shall submit to environmental protection bureau an environmental impact report compiled by an agency engaged in environmental impact assessment; if the project belongs to "intermediate cutting", it needs to submit an environmental impact form. With respect to desert prevention and control engineering, constructors shall submit to the environmental protection bureau an environmental impact report compiled by an agency engaged in environmental impact assessment.

For any proposed CDM A/R, if its purpose is to create a forest for public interests, constructors need to submit environmental impact registration form to environmental protection bureau, which will be reviewed and approved by the latter.

3.9 Biodiversity Survey

Project participants need to describe biodiversity status of the project area in PDD, including various natural reserves, precious or endangered species, and habitats under national or provincial protection, International Union for Conservation of Nature (IUCN) red list species (endangered and vulnerable species) and biodiversity information on afforestation/reforestation land.

Information on the biodiversity of project zone can be accessed by checking local literature, filed study, and semi-structured interview; information on project land plot can be obtained by baseline survey; using line transect or quadrat method to investigate species and distribution of fauna and flora, habitat, protection status, and threatening factors.

3.10 Design Afforestation

1. **Design method**

A project is designed in combination with satellite image analysis, on-site mapping, community village interview, and professional design work. Land eligibility results and identified project boundary is the basis of CDM A/R project design. Use project baseline measurement to investigate the land conditions, and adopt the participatory planning method to investigate community willingness during RPA. In light of natural condition and eco-development requirement, plus national technical regulations, afforestation management measures should be designed in line with the actual situation.

2. **Design principle**

CDM A/R project is designed following CDM modalities and procedures and the existing national laws and regulations, technical standards, or codes; it follows local conditions to plant trees on proper land, with full consideration of biodiversity protection and prohibit the selection of invasive plant species; it selects multiple tree species to avoid monoculture and create sustainable, stable, and multifunctional forest ecosystem; it gives full consideration to social and economic condition of the project area, respect the willingness of farmers (especially minority groups) and reserved areas, so that a project can benefit communities to the maximum and protect the rights and interests of communities concerned, and forest resource can be managed in a sustainable manner.

3. **Design basis**

CDM A/R project must follow national, local, or technical standards, procedures or management methods, such as Artificial Afforestation Technical Regulations (GB/T15776–2006), standard for Non-commercial Forest Construction (GB/T18337.1–2001, GB/T18337.2–2001, GB/T18337.3–2001), Design Code for Afforestation Operation (LY/T1607–2003), Regulation for Tending of Forest (GB/T15781–2009), Tree Seedling Quality Grading of Major Species for Afforestation (GB6000), Technical Regulation for Cultivation of Tree Seedlings (GB/T6001), Technical Regulations of Containerized Seedlings (LY1000), and Rules for Forest Tree Seed Testing (GB2772).

4. **Design highlights**

(1) Select tree species

Project participants need to fully respect community willingness and protect biodiversity. On the basis of that, they should follow the principle of planting trees on proper land in an economic and effective approach, trying to select local plant species and fast-growing species. Invasive species are prohibited.

(2) A/R land cleanup and soil preparation

It shall be clearly identified as to the land cleanup, method, scale, and time of the land preparation. In order to avoid or reduce carbon emission caused by A/R, cleanup and soil preparation should be down by man power, full cultivation, and controlled mountain fire are prohibited for preparation work.

(3) Planting measure

Participants need to draft detailed A/T plan by year and by tree species, including plantation mode, distribution mode, planting density, planting season, etc.

It is required to create mixed forests for ecological co-benefits or set up manmade tree groups with combined structure. Planting density should refer to national or local regulations on convention A/R, and should be determined in combination with the features, growth condition and forest carbon increase of tree species.

(4) Seedling cultivation

Design seedling source and breeding method, technology, and quality classification in accordance with project scale and planting plan.

(5) Young tree maintenance

In light of tree species, management method and time schedule should be designed to foster tree growth, including weeding, earth, seedling, debudding, replanting, irrigation, and riddance of disease species.

(6) Forest management

Covers young tree management, forest protection, and thinning.

References

NDRC (2012) Interim measures for the administration of voluntary greenhouse gas emission reduction transactions (No. 1668 [2012] of the NDRC). http://cdm.cccchina.gov.cn/Website/CDM/Upfile/Fice2894.pdf

UNFCCC (2003) Land use, land use change and forestry: definition and modalities for including afforestation and reforestation activities under article 12 of the Kyoto protocol in the first commitment period: draft consolidated text. FCCC/SBSTA/2003/4. http://unfccc.int/resource/docs/2003/sbsta/04.pdf

Reforestation Projects at Pearl River Basin of Guangxi in China

Sanzhong He and Zhuping Mo

Summary

The reforestation project at the Pearl River Basin of Guangxi in China is a bilateral project between the Department of Forestry of Guangxi Zhuang Autonomous Region and the BioCarbon Fund (BIOCF) of the World Bank. Under the leadership of Department of Forestry of Guangxi Zhuang Autonomous Region, the project area is 4,000 ha and is implemented at Huanjiang County and Cangwu County of Guangxi. The purpose is to obtain (measureable, reportable, and verifiable) quality CER by exploring CDM A/R technology and methodology. CER are sold to obtain carbon sink profit that can increase local people's income and contribute to local biodiversity and community development. Meanwhile, experience can be summed up through pilot project, so that CDM A/R can be implemented in China and provide pilots for other countries in developing similar projects.

In November 2005, "the baseline and monitoring methodology for reforestation on degraded land" developed by this project was approved by the CDM-EB and became the first CDM A/R baseline and monitoring methodology (AR-AM0001) globally. In November 2006, the project was officially registered with the CDM-EB and became the first reforestation project registered with the CDM-EB in the world.

In June 2006, Guangxi Huanjiang Xinghuan Yinglin Co., Ltd. (as the project developer) signed an emission reduction purchase agreement (ERPA) with BIOCF. The project is expected to generate 462,000 tCERs by 2017, which will be purchased by BIOCF at the unit price of US$4.35/ton.

The starting date of this project activity and its crediting period is April 1, 2006. The implementation period is 30 years, and the crediting period is 30 fixed years.

S. He (✉)
Guangxi Forestry Department, Foreign Cooperation Office, Guangxi, China
e-mail: hesanzhong@126.com

Z. Mo
Institute of Forest Planning, Guangxi, China
e-mail: mzp1968@163.com

Z. Lu et al. (eds.), *Forest Carbon Practices and Low Carbon Development in China*, https://doi.org/10.1007/978-981-13-7364-0_4

The project is cooperated by farmers (communities) and forest farms, joint afforestation by farmers or single-family afforestation, which reflects the characteristics of Guangxi's forest management.

The project has set up a sound operation, management, monitoring, and measurement system, which supervises the implementation of project boundaries, project activities, changes of carbon stocks in the project, emissions, and leakage generated from project activities. Strict quality assurance and quality control procedures have been drafted to ensure monitoring accuracy.

In April 2012, the Japan Audit and Certification Organization conducted on-site certification of the monitoring results. On December 27, 2012, the CDM-EB endorsed the certification report and formally issued 131,964 tons of tCERs.

1 Project Background

At the end of 2003, Department of Forestry of Guangxi Zhuang Autonomous Region cooperated with the World Bank for the first time to jointly operate "the comprehensive forestry development and protection project in Guangxi" with the World Bank loan. The project applied for a loan of US$100 million from the International Bank for Reconstruction and Development (IBRD) to build 3 million mu (200,000 ha.) of fast-growing and high-yield commercial timber forest. During the project review process, government agencies suggest seeking grants as support package to lower loan conditions. In order to facilitate a successful loan, the World Bank and Guangxi worked together to actively seek new sources of investment and explore new opportunities for project cooperation. At that moment, the World Bank was launching BIOCF program. As a result, a decision was made that Department of Forestry of Guangxi Zhuang Autonomous Region will take lead to apply for a pilot project from forest project under the framework of "the comprehensive forestry development and protection project in Guangxi", namely "the reforestation project at the Pearl River Basin of Guangxi in China". The project will be utilized to explore CDM reforestation technologies, methods, and carbon trading mechanisms. High-quality CERs will be obtained and sold to get carbon profit for increasing the income of local farmers and promoting the conservation of local biodiversity and communities' development. At the same time, by summing up the experience of pilot project, global CDM A/R project will be further implemented.

The project is jointly funded by government and international organizations. Government organizes afforestation agencies and farmers to work on common development. The forest land used by the project is all degraded land located in remote areas, and is also at ecologically vulnerable regions of the upstream of the Pearl River Basin. In order to encourage communities and rural households to plant trees in these areas, the government of Guangxi Zhuang Autonomous Region offered policy and technical support. The World Bank's BIOCF promised to buy the future forest carbons in the form of contract buyers. In this way, in addition to the expected revenue from forest products, the project implementation entities can also sell the forest carbon generated by the project to increase the income of landowners and mobilize the initiative of community farmers to participate in the project.

2 Project Progress

The preparatory work for "the reforestation project at the Pearl River Basin of Guangxi in China" has been underway since the beginning of 2004. It has undergone project identification, project baseline investigation, project concept document and carbon financing document drafting, development and application of methodology approval, PDD drafting, validation by the DOE, application to the National Development and Reform Commission (NDRC) for administrative approval for CDM project. Upon approval of the project, the project will register in the United Nations CDM-EB the project was also certified by the DOE in April 2012.

(1) Project identification (April 2004–October 2004)

According to the requirements of CDM A/R for land eligibility under the first commitment period of the Kyoto Protocol, the project area of proposed reforestation activities shall be physically investigated to collect natural, social and economic conditions of the project area, especially the information of land use, land coverage history, and current status. Project feasibility study will be carried out. Since there has never been any experience with such project before, during the project identification period, the World Bank's BIOCF has also sent its experts to participate in the selection of afforestation sites, with main focus on determining whether the project meets the requirements of land eligibility, and deciding on tree species and plantation model for the project.

(2) PDD Preparation (September 2004–February 2005)

Project document was prepared based on the specific requirements of the World Bank's BIOCF for the project, the purpose of which is to identify a real buyer of the CER generated from the project. In September 2004, the World Bank's loan of "Guangxi Integrated Forestry Development and Protection Project" was approved by the World Bank expert group and the project began the preparation. To this end, an expert group on forestry carbon project was set up, and local forestry technicians joined the group to conduct in-depth baseline investigation in project counties and villages (communities). The survey mainly includes: natural conditions, land resources, vegetation types, biological resources, forest management and conservation models, socioeconomic conditions and development plans, assessing the potential impact on local farmers, especially the impact on the economy and culture of ethnic minorities; the history and current status of biological species, as well as the impact of project activities on biological species and communities and the environment are also investigated; Participatory Rural Appraisal approach is adopted to conduct in-depth surveys of communities in order to understand the willingness of afforestation for farmers and afforestation entities. The afforestation plots, afforestation tree species, plantation models, and detailed management plan for the proposed CDM reforestation project are decided based on their willingness. The financial status, organizational implementation, and management capacities and experience of various entities involved in the project are also investigated. On December 30, 2004, documents preparation including carbon financing document, environmental and social evaluation reports, and project feasibility study report were completed and submitted to the Management Committee of World Bank's BIOCF and the funders for review. The World Bank approved the feasibility study report in mid-February 2005 and reached an initial agreement of intention on CER purchase. At the same time, the project was submitted to the National Forestry and Grassland Administration and the Office of National Leading Group to Address Climate Change, and a letter of no objection was issued by the latter.

(3) Development and Application for Approval Methodology (March 2005–November 2005)

According to the *CDM A/R Project Approaches and Procedures under the Kyoto Protocol*, CDM A/R project must be registered with the UN CDM-EB. In the PDD submitted for registration, the baseline methodologies and monitoring methodologies used for calculating the CERs generated from the project must be clarified, and such methodologies must be officially approved by the CDM-EB in advance. Since no approved methodology was available at that time, the project team followed the abovementioned *CDM A/R Project Approaches and Procedures* and the CDM-EB's regulations, referred to the IPCC's *Good Practice Guide for Land Use, Land-Use Change, and Forestry* (*LULUCF*), and draw upon related theories and research outcome of forest ecology and forest management over the years. With respect to the specific site conditions, survey and calculation of vegetation biomass and the results of investigations on the boundaries of afforestation plots, the baseline methodologies and monitoring methodologies applicable to the implementation of CDM reforestation projects on degraded land was submitted to the CDM-EB on June 14, 2005, namely the "Baseline Methodology and Monitoring Methodology for Degraded Land Reforestation". On November 25, 2005, the methodology was approved by the CDM-EB and became the first approved CDM A/R methodology in the world (AR-AM0001).

(4) Completion of PDD

According to the requirements of the CDM-EB, a field project pilot must be provided as a support case while applying for methodology. Therefore, the PDD preparation is actually together with the development and application of methodology. Since March 2005, the expert group drafted a project site survey and project design work manual, and completed the project site quality evaluation report at the end of April 2005 on the basis of the investigation of forest site, soil, and vegetation for the project. Subsequently, project afforestation design was carried out, afforestation land, afforestation tree species, afforestation model, and afforestation technology were further implemented, and the afforestation entity and the form of management after afforestation were clarified. By the end of December 2005, the investigation of the existing vegetation biomass in the afforestation site was completed. Project archives and GIS database were established, and the PDD preparation was completed.

(5) Project Validation and Application for project approval

From February to April of 2006, according to China's "Operation and Management Methods for Clean Development Mechanism Project", the project implementing entity formally submitted an approval application of CDM project to the NDRC, and a designated operational entity authorized by the CDM-EB—TUV was appointed to review the project. TUV project auditors went into the project area to cross-check with the description of afforestation sites in PDD and conducted on-site inspections on the eligibility and additionality of the afforestation sites. At the same time, they also went into communities where the project was located to understand the farmers' willingness to participate in the project, as well as the contract relationships with the afforestation entities. They learned about the actual living standards of farmers in the project area, whether they really participated in the project on voluntary basis, and whether the benefit distribution of wood and forest carbon is reasonable. They completely verified the contents described in PDD. In March 2006, the NDRC issued an approval letter for the project. In November 2006, the CDM-EB officially approved the project, making it the first CDM reforestation project that is successfully registered in the world.

(6) Project negotiation and signing of formal purchase agreement

In April 2006, the Guangxi Huanjiang Xinghuan Yinglin Co., Ltd. (as the project implementation entity) negotiated with the International Bank for Reconstruction and Development (as the custodian of the BIOCF) on purchasing the quantity of project emission reduction, and determined the price tCER generated by this project at US$4.35/ton. According to the PDD, the 462,000 tons of tCER generated by the project before 2017 would be purchased by BIOCF. On June 30, 2006, the two parties formally signed an emission reduction purchase agreement.

(7) Project Implementation

In April 2006, the "reforestation project at the Pearl River Basin of Guangxi in China" was officially started and entered the phase of full implementation. Implementation activities include seedling cultivation, clearing and soil preparation, seedlings, survival rate inspection, replanting, young forest nurturing, archiving and management, and project activities monitoring.

(8) Project monitoring

According to PDD and monitoring plan, the first monitoring of the quantity of project emission reductions will be conducted in 2012. Therefore, the Department of Forestry of Guangxi Zhuang Autonomous Region commissioned the Guangxi Forestry Inventory & Planning Institute to monitor the removal of net anthropogenic greenhouse gas removals by sinks between November 2011 and January 2012.

(9) Project Certification

During April 6–14, 2012, according to the CER purchase agreement signed with the BIOCF, experts from the Japan Audit and Certification Organization were hired to certify project monitoring. The Japan Audit and Certification Organization submitted a project certification report to the CDM-EB in August 2012. On December 27, 2012, the certification report was approved by the CDM-EB and 131,964 tons of tCERs was officially issued.

3 Project Content

3.1 Project Area

The project area selects Huanjiang County and Cangwu County, which are important ecological conservation zones and ecologically fragile areas in the upper stream of the Pearl River. It covers 14 villages in 6 townships of Huanjiang County and 13 villages in 4 townships of Cangwu County. The selected afforestation sites are heavily degraded and low-productivity wastelands. Since the 1950s, due to numerous large-scale deforestation, excessive exploitation, grazing, and fire, the land is continuously deteriorating. The location of the reforestation site is remote, with poor site conditions and unattractive economy. The project has a total afforestation area of 4,000 ha, with each County having 2,000 ha. Among the 2,000 ha in Huanjiang County, 830 ha are adjacent to Mulun National Nature Reserve and Jiuwanshan National Nature Reserve, 1,170 ha are located between the two Nature Reserves; the 2,000 ha of afforestation land in Cangwu County are all located in the Pearl River basin management area that suffers serious land loss and a fragile ecological condition.

3.2 Project Goal

The goal of the project is to explore CDM A/R technologies and methods and obtain high-quality CERs through project implementation. The forest carbon transaction revenue through the sale of CERs will be used to increase the income of local farmers and promote local biodiversity conservation and community development. At the same time, through the summarization of practical experiences in pilot projects, the implementation of CDM A/R projects in China will be promoted and serve as demonstrations for similar projects in other countries.

3.3 Implementing Entity

The implementing bodies that are directly involved in the project include Huanjiang Xinghuan Yinglin Co., Ltd., Huanjiang Lühuan Forestry Development Co., Ltd., Kangyuan Forest Farm of Cangwu County, Fuyuan Forest Farm of Cangwu County, 18 farmer households and 12 farmers.

Under the "Operation and Management Methods for Clean Development Mechanism Project", a project can only be declared by one body. In the above-mentioned project implementation entities, the Huanjiang Xinghuan Yinglin Co., Ltd., due to its strong capacity fundraising and active engagement in afforestation in the region, is elected as the representative of all project participants implementing the project. It took the responsibility for submitting the PDD to NDRC, negotiating with BIOCF, signing agreement, reporting the progress project implementation in the future and applying for the payments of forest carbon revenue. The remaining company, two tree farms, farmers, and farmer households signed commission agreement respectively with this company.

3.4 Operating Mode and Project Management

1. **Operating mode**

There are three forms of operation in this project, the first is that farmers (communities) and forest farms (companies) cooperate on afforestation on a sharing basis, that is, farm households or villagers groups provide land and labor, the two companies and forest farms provide funds and technology to cooperate, respectively. The forest products are divided into proportions, which are determined according to forest productivity and transportation conditions. This is the most important form of operation and accounts for 90% of the total afforestation area of the project. The other two forms are joint farm household afforestation and individual farmer afforestation.

The land used for the project, subject to lease or joint venture between company (forest farm) and the farmer household/community, is determined by signing a lease agreement or joint venture contract. All contracts are kept on record in the county forestry bureau.

2. **Project management**

In order to facilitate project development and implementation, Guangxi set up a complete management system, as shown in Fig. 1.

(1) Responsibilities of the project office of Department of Forestry of Guangxi Zhuang Autonomous Region

- Organize the daily management of project applications and implementation;
- Coordinating the relationship among all relevant parties;

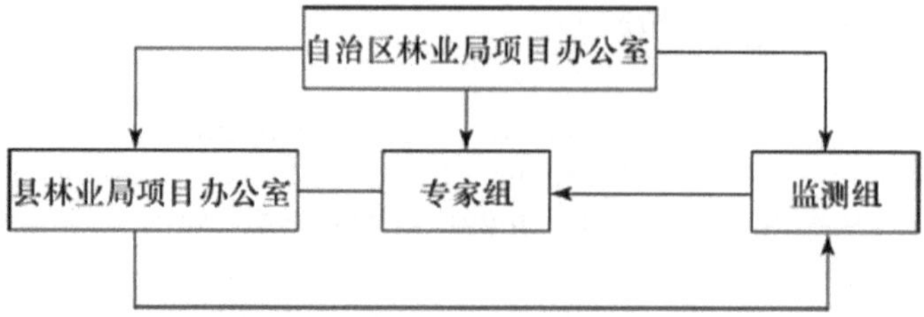

自治区林业局项目办公室	Project office of Department of Forestry of Guangxi Zhuang Autonomous Region
县林业局项目办公室	Project office of County Forestry Bureau
专家组	Expert team
监测组	Monitoring team

Fig. 1 Diagram of project management system

- Maintain contact and communication with the World Bank's BIOCF committee and national representatives;
- Examine and supervise project implementation;
- Organize provincial level training for county project offices and entities (farmers);
- Review project annual plan and annual report;
- Coordinate major technical and financial issues during project implementation;
- Organize monitoring and verification;
- Monitor quality assurance.

(2) Responsibilities of the Project Office of County Forestry Bureau

- Organize the implementation of on-site work of project;
- Check the project implementation progress;
- Prepare and implement project annual plans;
- Supervise and assist project entities (farmers) to prepare funds in a timely manner;
- Report on project progress;
- Provide technical training for forest farm (farmer household) organizations participating in the project;
- Provide technical guidance, inspection, and supervision for project activities;
- Examine the survival rate of afforestation in the first 3 years of the project;
- Organize technical personnel to cooperate with the monitoring team to perform field surveys;
- Archiving and management of project data and documents at county level (including farmers' groups);

- Communication between forest farms (companies) and communities (farmers);
- Forest carbon revenue management and distribution;
- Coordinate technical and financial issues during project implementation.

(3) Responsibilities of expert team

The expert team included experts from forestry, resource management and monitoring background within the Guangxi Zhuang Autonomous Region and also several national experts. Its specific responsibilities are:

- Develop inspection and verification standards, operating procedures, monitoring manuals, training materials, and forest growth standards for the project;
- Provide technical training;
- Project technical consultation;
- Verification of survival rate and quality control;
- Quality control and quality assurance of on-site measurement results of monitoring team and verification teams;
- Prepare a monitoring report.

(4) Responsibilities of monitoring team

The monitoring team included relevant experts and professional technicians from the Guangxi Forestry Inventory & Planning Institute. Their specific responsibilities are:

- Regular monitoring within project boundaries;
- Project stratification;
- Prepare and implement sample designs;
- Prepare *Standard Operating Procedures* through field measurement;
- Preliminary analysis of field measurement data;
- Assist experts to analyze the collected data;
- Assist expert team to prepare monitoring report.

3.5 Project Timeline

The start date of project activity and its crediting period is April 1, 2006. The proposed project activity has a fixed period of 30 years and cannot be extended.

3.6 Land Eligibility

The land eligibility used in this project was confirmed by investigations in three aspects. The field survey conducted by expert team showed that the project site was a heavily degraded, low-productivity wasteland, covered only by herbs and a few shrubs. It is neither forest nor belonging to any temporary forestless land. Although there are a small number of trees on a small area of land, canopy density is much lower than 20% (forest canopy density's threshold defined by China). According to the inventory of forest resources conducted by China every five years and the recent land use maps and forest coverage maps, the project's forest land is a wasteland. An interview with local farmers and communities on the historical changes in the land use (land coverage) of the project afforestation sites showed that the project site has been forestless wasteland since 1989. Therefore, it fully meets the eligibility requirement on CDM reforestation project for the selected site.

3.7 Baseline Scenario

The baseline scenario of the project has two levels: qualitative baseline of forest land; and baseline of existing non-tree vegetation biomass in forest land. As far as the qualitative baseline is concerned, the project site used to be a forest land in the 1950s. However, three major deforestation activities happened between the 1950s and the 1980s, which resulted in the gradual degradation of forest land. The first happened in the late 1950s during the "Great Leap Forward" movement and the "Great Iron and Steel Making" campaign. Large areas of forest were logged for steelmaking. The second occurred during the "Cultural Revolution" (1966–1976), when most of the remaining secondary forest was developed as agricultural land. The third occurred in the late 1970s and early 1980s. Due to the reform of forestland ownership, the land, including the remaining secondary forest, was allocated to individual farmers. Farmers lack confidence in the reform and immediately logged down the forest without any renewal measures. By mid-1980s, these lands were all forestless. However, human disturbance in these lands did not stop there, such as excessive logging for fuelwood and excessive grazing. At the same time, fire frequently occurs on agricultural lands, which poses a great threat to the lands. As a result, the land will continue to degrade at baseline, and land loss will become more serious, resulting in continued reduction of biomass and soil carbon reserve.

Interviews with local farmers also showed that they face many obstacles planting trees on these lands, including low land productivity, lack of upfront investment, difficulties in obtaining loans for reforestation, lack of technology, and lack of marketing capabilities.

As for the baseline of existing non-tree vegetation biomass, in April 2005, experts from the Guangxi Forestry Inventory & Planning Institute, Forestry College of Guangxi University, and the Forest Ecology and Environment Research Institute of the Chinese Academy of Forestry conducted an on-site survey to compile a classified evaluation report. According to the evaluation results, the project site was divided into seven types. Typical sampling selection method was used. The sample setup covered all vegetation types, and a typical sample plot was chosen for the vegetation type. After the sample is set up, its position was first determined by GPS and the coordinates (latitude, longitude and altitude) were recorded. The area of herbaceous sample is 1 m × 1 m, and the area of shrubs is 2 m × 2 m or 2 m × 1 m (depending on the size and height of shrub). A total of 110 sample plots were set up for baseline investigations of non-tree vegetation biomass.

The result of summarizing non-tree vegetation biomass per unit area of site type is as follows: the four site types in the project area of Huanjiang County are mainly herbaceous, with biomass accounts for more than 80% of the total. The biomass of different types of shrubs is close to each other, with about 3 tons of biomass per hectare, of which aboveground biomass is about 60–70% and belowground about 30–40%. There are three types of shrubs in the project area of Cangwu County, which enjoy various shrub types and large biomass, with about 5 tons per hectare.

3.8 Additionality

Using the "comprehensive evaluation tool for identifying the baseline scenario and additionality of CDM A/R projects" approved by the CDM-EB, the proposed CDM project activity has the additionality.

First, after a preliminary screening of the proposed project based on starting date of project activity, it has been identified that the alternative land use plan that is in line with current laws and regulations is to maintain the current state of degradation.

Second, project investment analysis proves that the project is not economically attractive without the benefit of forest carbon transactions. We chose the benchmarking method and used the financial analysis tool from project concept book developed by the World Bank's BIOCF to analyze the project's financial internal rate of return (FIRR) under the condition of w/o carbon income. Analysis shows that, with 30-year crediting period, and in the case of without/with carbon income, the internal rate of return for the project is 8.53% and 15.02% respectively (calculated at US$4/ton of CO_2 equivalent, approximately equivalent to the purchasing price of BIOCF). In other words, only when there is carbon income can FIRR exceed the benchmark yield (i_c, 12%), and the project will be economically attractive. This proves that this project is not economically attractive without carbon income.

Sensitivity analysis also shows that in the absence of carbon revenue, FIRR is lower than the benchmark yield (i_c, 12%) when construction cost falls by 10%, or forest products or price rises by 10%, or operating cost decreases by 10%; and taking into account the carbon revenue, FIRR is higher than the benchmark yield (i_c, 12%) when construction cost increases by 10%, or forest products or price drops by 10%, or operating cost increases by 10%. The analysis shows that when the main parameter changes by 10%, the project is not economically attractive without carbon revenue, and it will be economic attractiveness when carbon revenue is available.

This is followed by obstacle analysis. In terms of investment, technology, institution, and market risk, without the project, afforestation on the selected project site may encounter obstacles. Because of the above obstacles, without the involvement of carbon trading, the project land can't have any afforestation activities and will remain as it is now, namely continuous land degradation, accompanied by grazing, farming, logging, and frequent fire burning.

Finally, analysis of the impact of registered CDM. Approval and registration of proposed CDM project activities can reduce the identified obstacles in economic and financial difficulties, enable project activities to be implemented, and can produce the following benefits: Absorb the carbon in the atmosphere and obtain carbon income, reduce predictable investment risks in project activities; establish closer links between farmers, communities, forest farms and governments, strengthen technical training, production services, build product marketing networks, and improve farmers' afforestation skills and project management capabilities.

The above analysis shows that during the implementation of CDM reforestation projects in the project area, there are obstacles in terms of capital, technology, institution, and market risk in the absence of carbon revenue, and the FIRR of the project is lower than the benchmark set by the Chinese government; after the project is registered as a CDM A/R project, the resulting carbon sink revenue and incentives can overcome the above-mentioned barriers and allow the project to be implemented. This confirms the additionality of the proposed CDM A/R project activity.

3.9 Permanence

Because the carbon fixed in the forests cannot be forever secured, it is not permanent. A solution is needed at certain time in the future to tackle this problem. The project uses the issued temporary CERs (tCERs) to deal with the non-permanence.

3.10 Estimate of Net Anthropogenic GHG Removals by Sinks of the Project

Net anthropogenic greenhouse gas removals by sinks = Actual net GHG removals by sinks—Baseline net GHG removals by sinks—Leakage from GHG emission in total.

1. **Estimate of actual net GHG removals by sinks**

The estimate of actual net GHG removals by sinks includes direct N_2O emissions due to nitrogen inputs, reductions in carbon in existing vegetation, and changes in carbon of aboveground and belowground biomass.

Use local growth equations to estimate the changes in biomass caused by CDM project activities. These equations come from China's forestry inventory data, which has been conducted once every 5 years since the 1970s. Growth data (volume per hectare) were converted to biomass using wood density, BEF and root/shoot ratio and using Eqs. (6) and (9) described in Sect. 5 of the Approved Methodology AR-AM0001. These parameters come from the LULUCF GHG Inventory Report of the *Initial National Notification of China.*

Reforestation was carried out in 2006 and 2007, respectively, and the carbon sequestration of each reforestation species was calculated separately for different years and then accumulated together.

The N_2O emission caused by nitrogen fertilizers is estimated by using the formula (23, 25) in Part 7 of the approved methodologies AR-AM0001 and using the default values of GPG 2000 and 1996 IPCC guidelines. For fertilizers used in the reforestation of *eucalyptus*, the application rates per tree are 75 gN, 30 gN, and 40 gN, respectively, in the first, second, and third year of plantation or renewal.

The actual net GHG removals by sinks are the change in carbon stock in aboveground and belowground biomass, minus the N_2O emission caused by nitrogen fertilizer. Carbon stock of existing non-tree vegetation biomass is set at 5 tons of carbon per hectare (based on the default value of GPGLULUCF live grassland biomass) and is deducted from carbon stock in the first year of crediting period. Details of the 30-year period are shown in Table 1.[1]

2. **Estimate of benchmark net GHG removals by sinks**

To use the conservative estimation, in the baseline scenario, the change in carbon stock of above and belowground biomass is set at zero for land without tree growth; by predicting the growth of existing trees, it is estimated that there will be biomass change on a small number of land with trees. This serves as the baseline for the project's net GHG removals by sinks.

There are a small number of trees on a land of 35 ha in the project area. There are on average 40 trees per hectare with an average diameter of 2.89 cm and an average height of 3.42 m. The main tree species are *liquidambar formosana*, *Elaeocarpus sylvestris* and *Albizia julibrissin Durazz.* The average tree age is about 6 years. Changes in the carbon reserve of above and belowground biomass of the existing trees are shown in Table 1. Formula (6, 9) of Part 5 of the Approved Methodology AR-AM0001 and the growth curve of local hardwood tree are estimated. Details of 30-year period are shown in Table 2.

3. **Leakage estimate**

The proposed CDM project uses all local labor, hence there is no need for labor transportation. In addition to *Eucalyptus*, all seedlings will be grown locally and sent to the reforestation area by people. Moreover, wood will be harvested after 7 years, and resin harvested after 16–20 years. Mid-range diesel vehicle is the often

[1]Notice on Conducting Pilot Projects of Carbon Emission Trade by the General Office of NDRC (NDRC [2011] No. 2601) http://www.sdpc.gov.cn/zcfb/zcfbtz/2011tz/t20120113-456506.htm.

Table 1 Actual net GHG removals by sinks

Year no.	Year	Aggregate carbon stock change/tons of CO_2		Annual GHG emission/(tons of CO_2 equivalent year^{-1})	Aggregate GHG emission/tons of CO_2 equivalent	Annual actual net GHG removal by sinks/(tons of CO_2 equivalent·year^{-1})	Aggregate actual net GHG removals by sinks/tons of CO_2 equivalent
		Aboveground biomass	Belowground biomass				
1	2006	−24,934	−15,276	−437	−437	−40,647	−40,647
2	2007	18,494	−4022	−252	−688	54,430	13,783
3	2008	64,612	7982	−264	−952	57,858	71,642
4	2009	110,238	19,926	−41	−993	57,530	129,171
5	2010	159,105	32,858	0	−993	61,799	190,970
6	2011	216,241	48,004	0	−993	72,281	263,251
7	2012	277,519	64,000	0	−993	77,275	340,526
8	2013	333,734	78,131	0	−993	70,345	410,872
9	2014	310,180	71,838	−57	−1050	−29,904	380,968
10	2015	303,050	71,348	−126	−1176	−7746	373,221
11	2016	298,113	71,523	−217	−1393	−4978	368,243
12	2017	375,795	92,250	−160	−1553	98,249	466,493
13	2018	438,262	108,185	−91	−1644	78,310	544,803
14	2019	471,587	114,662	0	−1644	39,802	584,605
15	2020	492,464	117,362	0	−1644	23,577	608,181
16	2021	550,902	133,268	0	−1644	74,345	682,526

(continued)

Table 1 (continued)

Year no.	Year	Aggregate carbon stock change/tons of CO_2		Annual GHG emission/(tons of CO_2 equivalent year^{-1})	Aggregate GHG emission/tons of CO_2 equivalent	Annual actual net GHG removal by sinks/(tons of CO_2 equivalent·year^{-1})	Aggregate actual net GHG removals by sinks/tons of CO_2 equivalent
		Aboveground biomass	Belowground biomass				
17	2022	499,288	119,859	−57	−1701	−65,080	617,446
18	2023	432,422	102,717	−126	−1827	−84,133	533,313
19	2024	359,582	83,136	−217	−2044	−92,639	440,674
20	2025	395,331	90,537	−160	−2204	42,991	483,665
21	2026	421,019	94,769	−91	−2295	29,828	513,493
22	2027	485,293	112,457	0	−2295	81,962	595,455
23	2028	546,792	129,564	0	−2295	78,605	674,061
24	2029	603,039	145,239	0	−2295	71,922	745,983
25	2030	584,768	141,276	−57	−2352	−22,291	723,692
26	2031	525,173	124,788	−126	−2478	−76,207	647,484
27	2032	462,397	106,821	−217	−2694	−80,961	566,523
28	2033	524,464	123,860	−160	−2854	78,947	645,470
29	2034	585,122	140,651	−91	−2946	77,357	722,827
30	2035	641,035	156,136	0	−2946	71,398	794,225

Table 2 Baseline net GHG removals by sinks

Year no.	Year	Annual baseline net GHG removals by sinks/(tons of CO_2 year^{-1})		Total	Aggregate baseline net GHG removals by sinks/tons of CO_2		Total
		Aboveground biomass	Belowground biomass		Aboveground biomass	Belowground biomass	
1	2006	6.1	1.7	7.8	6.1	1.7	7.8
2	2007	6.7	1.9	8.7	12.8	3.6	16.5
3	2008	7.4	2.1	9.5	20.2	5.7	25.9
4	2009	8.0	2.3	10.2	28.2	8.0	36.2
5	2010	8.6	2.4	11.0	36.8	10.4	47.2
6	2011	9.2	2.6	11.7	45.9	13.0	58.9
7	2012	9.7	2.7	12.5	55.6	15.7	71.4
8	2013	10.3	2.9	13.2	65.9	18.6	84.5
9	2014	10.8	3.1	13.9	76.7	21.7	98.4
10	2015	11.3	3.2	14.5	88.0	24.9	112.9
11	2016	11.8	3.4	15.2	99.9	28.3	128.1
12	2017	12.4	3.5	15.9	112.2	31.8	144.0
13	2018	12.9	3.6	16.5	125.1	35.4	160.5
14	2019	13.3	3.8	17.1	138.4	39.2	177.6
15	2020	13.8	3.9	17.7	152.3	43.1	195.4
16	2021	14.3	4.0	18.4	166.6	47.1	213.7
17	2022	14.8	4.2	19.0	181.4	51.3	232.7
18	2023	15.2	4.3	19.6	196.6	55.6	252.2
19	2024	15.7	4.4	20.2	212.3	60.1	272.4
20	2025	16.2	4.6	20.7	228.5	64.7	293.1
21	2026	16.6	4.7	21.3	245.1	69.4	314.4
22	2027	17.1	4.8	21.9	262.1	74.2	336.3
23	2028	17.5	5.0	22.4	279.6	79.1	358.8
24	2029	17.9	5.1	23.0	297.6	84.2	381.8
25	2030	18.4	5.2	23.6	315.9	89.4	405.3
26	2031	18.8	5.3	24.1	334.7	94.7	429.4
27	2032	19.2	5.4	24.6	353.9	100.2	454.1
28	2033	19.6	5.6	25.2	373.5	105.7	479.3
29	2034	20.0	5.7	25.7	393.6	111.4	505.0
30	2035	20.5	5.8	26.2	414.0	117.2	531.2

used transportation tool in the project area. According to the demand of *Eucalyptus seedlings*, the predicted products of woods and resin, estimation of fertilizer quantity used in reforestation area, formula (27, 28) of Part 8 of the Approved Methodology AR-AM0001 is used to estimate the leakage caused by vehicle use. The emission factor of diesel vehicle is 2.6353 kg CO_2/L (data from China's Initial National communication on Climate Change). The estimated leakage until 2012 is 232 tons of CO_2 equivalent. The figure will be 19,852 tons of CO_2 equivalent by the end of the crediting period. Details are shown in Table 3.

Leakage calculation is based on leakage monitoring logbook. In order to achieve accurate calculation of leakage, a leakage monitoring logbook is prepared in the PDD and recorded time and content are clearly specified. As a part of the project file, the logbook is kept in the archive rooms of project office of county forestry bureau and afforestation entities (company or forest farm). Daily record is jointly completely by technicians of county forestry bureau and forest farm (company). Shift monitoring card and leakage monitoring logbook must be recorded on a timely basis to avoid omission and mistake. The monitoring card and leakage monitoring logbook must correspond to each other without conflict.

The contents of leakage monitoring log mainly include: afforestation and forest management activities such as site preparation, planting, tending, fertilizing, prevention and control of forest pests and diseases, and forest fire prevention; migrant workers and management technicians engaged in logging and renewal at all levels (including province, county and owner); CO_2 emission (leakage) due to fuel burning of vehicles in the process of transporting seed. The specific monitoring indicators include start and end location, distances, types and quantities of transport vehicles, fuel consumption, fuel types, etc.

4. **Estimate of net anthropogenic GHG removals by sinks**

During the selected 30-year crediting period (April 1, 2006–March 31, 2036), the estimated net anthropogenic GHG removals by sinks generated by this project activity are approximately 770,000 tons of CO_2 equivalent. Detailed data is shown in Table 4.

3.11 Major Tree Species and Afforestation Patterns

With the Participatory Rural Appraisal approach, the interviews with local farmers and companies were done. On the basis of full respect for the willingness of local communities and farmers, and taking into account carbon sequestration, biodiversity conservation, soil erosion control and the value of related forest products, identified species include: *Pinusmassoniana, Liquidambar formosana, Cunning Lanceolata, Schima superba, Eucalyptus* and *Quercusgriffinith, Q. acutissima.* In addition to *Eucalyptus,* others are local species. Even though *Eucalyptus* was introduced into China as a garden tree species since 1889, after more than 100 years of adjustment, it has been widely planted in the southern regions including Guangxi, and has no indication of invasive species. *Eucalyptus* is included in this

Table 3 Leakage's estimate

Year no.	Year	Annual leakage (tons of CO_2 equivalent $year^{-1}$)	Aggregate leakage/tons of CO_2 equivalent
1	2006	−140	−140
2	2007	−27	−168
3	2008	−38	−206
4	2009	−6	−212
5	2010	0	−212
6	2011	0	−212
7	2012	−21	−232
8	2013	−83	−315
9	2014	−1376	−1691
10	2015	−1304	−2995
11	2016	−1317	−4312
12	2017	−23	−4335
13	2018	−105	−4440
14	2019	−367	−4808
15	2020	−367	−5175
16	2021	−46	−5221
17	2022	−2903	−8124
18	2023	−2903	−11,027
19	2024	−3009	−14,036
20	2025	−547	−14,583
21	2026	−491	−15,074
22	2027	0	−15,074
23	2028	0	−15,074
24	2029	0	−15,074
25	2030	−1386	−16,460
26	2031	−1671	−18,131
27	2032	−1685	−19,816
28	2033	−23	−19,839
29	2034	−13	−19,852
30	2035	0	−19,852

Table 4 Estimate of net anthropogenic GHG removals by sinks

Year no.	Year	Annual net anthropogenic GHG removals by sinks/(tons of CO_2 equivalent $year^{-1}$)	Aggregate net anthropogenic GHG removals by sinks/tons of CO_2 equivalent
1	2006	−40,795	−40,795
2	2007	54,394	13,599
3	2008	57,811	71,410
4	2009	57,514	128,924
5	2010	61,788	190,711
6	2011	72,269	262,981
7	2012	77,242	340,223
8	2013	70,250	410,472
9	2014	−31,294	379,178
10	2015	−9065	370,113
11	2016	−6310	363,803
12	2017	98,210	462,013
13	2018	78,189	540,202
14	2019	39,417	579,619
15	2020	23,192	602,811
16	2021	74,280	677,091
17	2022	−68,002	609,089
18	2023	−87,056	522,033
19	2024	−95,668	426,366
20	2025	42,423	468,789
21	2026	29,316	498,105
22	2027	81,940	580,045
23	2028	78,583	658,628
24	2029	71,899	730,527
25	2030	−23,701	706,826
26	2031	−77,903	628,924
27	2032	−82,670	546,254
28	2033	78,899	625,152
29	2034	77,319	702,471
30	2035	71,371	773,842

project because it can generate more CERs in the early stages of project crediting period and is favored by local communities, while other tree species grow relatively slow during the first few years.

Afforestation species/patterns include mixed forest of *Pinusmassoniana* and *Liquidambar formosana* (1050 ha), mixed forest of *Cunning Lanceolata* and *Liquidambar formosana* (450 ha), pure *Eucalyptus* forest (1000 ha), mixed forest of *Pinusmassoniana* and *Quercusgriffinith*, *Q. acutissima* (900 ha), mixed forest of *Pinusmassoniana* and *Schima superba* (600 ha).

In order to ensure that project implementation does not pose a negative impact on the environment, the *Regulations on Environmental Protection of plantation Forestry Management* was prepared at the project design stage. All the forestry activities, including land preparation, planting and tending are carried out along the contour line. Attention is paid to protecting natural vegetation and habitats of endangered fauna and flora.

3.12 Investment Overview

Project crediting period is for 30 years (i.e., 2006–2035). The estimated budget for the project is US$2.822 million (based on the exchange rate in 2006, this is equal to RMB22.83 million). The actual afforestation area was 3008.8 ha and the completed investment was RMB19.593 million, accounting for 85.8% of the estimated investment.

3.13 Key Elements for Monitoring

Project archive is set up in the project office of county forestry bureau and afforestation entity (forest farm or company), and personnel are designated to handle archiving and management. The archives include: shift monitoring card, leakage monitoring log sheet, PDD, technical data, signed contract, annual inspection and acceptance report, pictures, gallery, and database. Archives must be complete, with clear classification and in standard order.

(1) Monitoring project implementation

In order to ensure that the project is implemented in accordance with the technical and economic indicators in the PDD, shift monitoring cards and leakage monitoring logbook are designed, and the project is implemented at any time by record.

Due to the delay in afforestation time, the monitoring system established for forest growth of the project and for calculating net anthropogenic GHG removals by sinks did not start operation until November 2011. Based on CDM guidance document on forestry carbon projects, Guangxi Forestry Inventory & Planning Institute set up a monitoring team consisting of qualified and experienced experts in forestry resources inventory, and compiled the *Operating Procedures for the Reforestation Project at the Pearl River Basin of Guangxi in China.*

(2) Calculation of stratification and sample plot quantity before project implementation

In order to be easy for future monitoring, during project design, the project was stratified before implementation. In light of the differences in site type and forestation year, 12 strata were divided and 110 monitoring sites were designed.

During project implementation, due to the great changes in afforestation model and afforestation time of tree species, project implementation was stratified before monitoring and divided into 22 strata. Then, the *Calculation Tool for Monitoring Site Number of CDM A/R Project* (version 02.1.0) was used to recalculate the number of sample plots at each level. In order to ensure the independence of statistics at each level, at least 3 sample plots are set for each level, and a total of 102 sample plots are set (Table 5).

(3) Layout of sample plots

In order to avoid arbitrary selection of sample plots (such as sample plot center, sample plot reference point, horizontal movement of plot center), and ensure that sample plot is distributed as evenly as possible to each stratum, the location of fixed sample plots should be set systematically, but the starting point is random. The specific layout is as follows:

Step 1: Using geographic information system software (ArcGIS), the project area is divided into several grids with a spacing of 100 m × 100 m. Calculate the number of grid intersections that fall into each layer (the number of possible plot centers). The number of grid points in each layer starts from 1.

Step 2: In an Excel spreadsheet, a formula is used to generate a random number, where n represents the total number of intersections in each layer. The generated random number coincides with the grid intersection number. The random point is the center of the first plot.

Step 3: Starting from the center point of the first plot along the fixed direction of "West–East–South–North", the next plot is selected at a fixed interval. The fixed interval depends on the total number of grid intersections and plots within the carbon layer.

The sample plot is 400 m^2 (20 m × 20 m). However, if the shortest distance from the edge of sample plot to the project boundary is less than 10 m, or if part of the plot falls outside the project carbon layer or project boundary, the sample plot should be moved horizontally to the center of the plot.

Following field measurement, if the accuracy is less than 10%, the number of sample plots should be recalculated by using the above method, based on the standard deviation of biomass to be measured. Additional sample plots should be extracted and laid out using the method above.

Table 5 Table of levels after-project implementation

Tree species and stand type	Afforestation year	Level No.	Area/hectare	Number of sample plot	Note
Pinusmassoniana (Cangwu County)	2006–2007	S-1	369.4	7	
	2008	S-2	236.2	5	Immeasurable
	2009–2011	S-3	122.9	4	Immeasurable
Pinusmassoniana (Huanjiang County)	2007–2008	S-4	417.4	7	
	2008	S-5	165.3	4	Poor condition (high altitude) Immeasurable
	2009–2011	S-6	153.6	4	Immeasurable
Cunning Lanceolata	2006–2007	S-7	37.8	3	
	2008	S-8	94.4	4	
	2011	S-9	16.1	3	Immeasurable
Schima superba	2006–2007	S-10	152.6	4	Immeasurable
	2008–2009	S-11	119.6	4	Immeasurable
	2011	S-12	25.4	4	Immeasurable
Eucalyptus	2006–2007	S-13	518.0	3	
	2008–2009	S-14	63.3	19	
	2008	S-15	120.3	3	Affected by 2011 rain, snow and frost disaster
	2010–2011	S-16	120.1	6	
Liquidambar formosana	2006–2007	S-17	30.9	4	
	2008	S-18	58.5	3	
Quercusgriffinith, Q. acutissima	2008–2009	S-19	26.8	3	Immeasurable
Pinusmassoniana + Liquidambar formosana Irregular afforestation	2006	S-20	25.7	3	
	2008	S-21	83.5	3	
Pinusmassoniana + Cunning Lanceolata Irregular afforestation	2008	S-2	51.0	3	Immeasurable
Total			3008.8	102	

Note Some sample plots in certain levels are still at the lower end of measurable limit due to slow growth or poor site conditions, thus no measurement has been made

(4) Field measurement

GPS is used in the field measurement to determine the geographic location (GPS coordinates) of the sample plots. Then the plot location and its respective level are recorded and saved. After the four corners of the plot are determined, permanent markers (PVC pipe 5 cm in diameter and 30 cm in length, 20 cm vertically buried in soil and 10 cm exposed) are embedded in the center and four corners of the plot, so that they can be identified precisely for the next time monitoring.

(5) Measurement of tree DBH and tree height in sample plots

The minimum value determined by tree DBH is 2.0 cm. If more than two-thirds or more of the trees in the sample plot do not reach a DBH of 2.0 cm, this plot does not need to be measured and the value of biomass carbon reserve is set to 0; otherwise, for each tree with DBH reaching or exceeding 2.0 cm within the sample plot, the height and DBH need to be measured. The DBH and tree height measured in the field are recorded in the monitoring questionnaire.

(6) Frequency of monitoring

The monitoring frequency of carbon stock change of the project is once every 5 years.

4 Project Implementation

4.1 Main Activity

The Estimate of actual net GHG removals by sinks of project, guidance documents developing Participatory Rural Appraisal and organizing technical training.

1. **Preparation of project guidance documents**

The CDM reforestation project is a brand-new project with a particularly high demand on policy, technology, and social and environmental impact. To this end, during project preparation, a series of guidance documents and technical regulations were drafted, such as project publicity materials, participatory project design guidelines, environmental regulations for artificial forest management, and integrated pest management plans, so that the project can be implemented in accordance with rules and proceeded step by step.

2. **Development of Participatory Rural Appraisal**

The CDM reforestation project is not only an afforestation project responding to climate change, but also has relevance to promoting community development and biodiversity conservation. It utilizes both remote and degraded land, but the landowners are relatively disadvantaged groups of poverty and ethnic minorities. In order to maximize the expected goals of the project, local communities and people need to actively participate on a voluntary basis. Therefore, in the whole process of afforestation design and implementation, the project is carried out in a participatory manner. The (autonomous) district and county project offices organize task forces to hold village meetings in the proposed project area, distribute project publicity materials, and hold discussions with the heads of local village (prefecture), elders and villagers to understand their preferences, wishes, and concerns. The project engages local communities and farmers from the outset. The project design reflects their expectation on the community; they are able to benefit from participating in labor work of the project at the initial stages of project implementation.

3. **Organization of technical training**

During project implementation, the autonomous region organized 7 training sessions in total, with 620 people being trained; 8 training sessions were held at county levels, with 2,030 people attending these courses. In addition to lecture course training, a large amount of training is conducted at community and afforestation sites. Meanwhile, in order to maximize the environmental value of the project, reduce GHG emissions, and protect the existing carbon stock, during the implementation, controlled mountain fire and soil reclamation and preparation are strictly prohibited. Land is prepared along the contour line (diameter 40–50 cm, depth 40 cm) to minimize the damage to land surface, and to reduce the natural risk (fire, pests and diseases). In some areas, such as Huanjiang County, a model of afforestation with mixed tree species was also adopted.

However, due to continuous bad weather since project implementation, in the spring of 2008, the project suffered unprecedented rain, frost, and snow. The continuous drought from 2009 to 2011 caused a serious delay in the implementation. Only 3008.8 ha of afforestation was completed, which was 75.2% of total afforestation area that had been designed.

4.2 *Difficulty and Solution*

As mentioned above, the project originally designed 4,000 ha of afforestation area, of which only 3008.8 ha was completed actually; the original plan afforestation should have been completed in 2006 and 2007, but it was delayed until 2011 due to bad weather. In addition to natural disasters, the project also encountered lots of difficulties: the site conditions for some sites were rather bad, which is mainly reflected in high altitude, steep slopes, and high content of gravel, plus that the site is on the top of mountain hill with high wind, leading to low survival rate and poor growth. Therefore, continuous replanting or reforestation is needed, making the afforestation time extended, and there were still some forest lands failing to reforest after several times. In some places, farmers have poor awareness on contract and are arbitrary in their discretion. Especially after the reform of collective forest rights system in rural areas, the ownership of some project land was changed. The land originally provided by community in association with forest farms or companies no longer belongs to collective but owned by households. Some rural households violated their contracts or raised some additional demand, for example, requesting that forestry products dividend be increased, requiring company to help them build roads. If these requirements are not met, farmers will no longer provide land for cooperative forestation. Unexpected land disputes occurred in some places. The land used for the project was mostly degraded land in remote areas. There was no economic value. Adjacent farmers did not care about the use of these lands. After project implementation, farmers see that the project can generate benefits, and they compete for ownership of land use, creating disputes and delaying project implementation.

To deal with above difficulties, the project office of Department of Forestry of Guangxi Zhuang Autonomous Region, Guangxi Forestry Inventory & Planning Institute and the Forestry Bureaus of Huanjiang County and Cangwu County, as well as the governments of these two counties, conducted large amount of targeted and meticulous work: First, an expert group is formed to conduct on-site survey in areas with poor site conditions and poor growth of forest trees to analyze the cause of the problem. The main reasons for afforestation failure are: the provenance of the selected afforestation tree species was not suitable; the standard of seedlings selection was not high enough; some seedlings purchased from Hunan had unclear provenance. Therefore, the expert group guided the project implementation units to adopt more suitable source seedlings, while strictly controlling the quality of nursery. Reforestation is performed again and the quality of the project was improved. Second, strengthen communication with farmers breaching the contract. With respect to helping villages to build roads, as long as such requests are not too far off, for example, if the rented area or area of cooperative afforestation is larger, and the intended village roads are not very long, the company (forestry farm) will increase investment as much as possible, which is combined with poverty alleviation projects in the county to meet the requirements of farmers, or they help villages to build power supply and water supply facilities, and solve the problem in some project sites. Third, in conjunction with the reform of rural collective forest rights system, villages, or peasant households that have disputes over a landownership are coordinated. The involved parties are summoned calmly talk about policies, give reasons, trace back to history, and solve conflict over certain land, thus increasing the proportion of afforestation projects.

Despite the great efforts made by governments, forest management authorities, afforestation companies or forest farms at all levels, it is still not possible to complete all afforestation as designed. Because according to the rules of CDM, after the PDD is certified, the afforestation sites cannot be changed or moved. Therefore, the afforestation land designed for the project cannot be supplemented whatever the reasons for loss might be, and the project area can only be reduced. The only direction is to control the reduced project area within an acceptable range.

4.3 Payment and Distribution of Carbon Revenue

The actual net GHG removals by sinks of the project became positive since 2008. By the end of 2012, based on the certification results in April of that year, a cumulative amount of 131,964 tons of net anthropogenic GHG removals by sinks has been approved and certified by the CDM-EB. The application for carbon revenue payment was started since 2009, and will be paid in accordance with the carbon as expected in PDD. Currently, 4 applications have been made to the BIOCF (by year), with a total payment of US$570,207.

As the implementing entity of the project, Huanjiang Xinghuan Yinglin Co., Ltd. does not have foreign account, thus the carbon revenues paid by the BIOCF is first allocated to the special account of the project office of Department of Forestry of Guangxi Zhuang Autonomous Region, which is further allocated to the Forestry Bureaus of Huanjiang County and Cangwu County, and the County Forestry Bureau according to the afforestation areas that have been completed or reached project standard by the implementing entities and farmers. County forestry bureaus pay the implementing entities and farmers accordingly.

The basic unit of carbon revenue is afforestation area, because in spite of the great differences in the growth of tree species, and the resulting CERs are different. Take *eucalyptus* as an example. Its early growth is several times higher than that of *Schima superba* or *Quercusgriffinith, Q. acutissima.* But before monitoring, it is very difficult to allocate the carbon sink revenue according to the growth amount, biomass, and CER generated by each tree species.

4.4 Project Certification

Pursuant to the project design and the carbon purchase agreement, the experts from the Japan Audit and Certification Organization conducted the first certification of the project from April 6–14, 2012.

The purpose of certification is as follows: to confirm that the project activities are implemented in accordance with the registered PDD; check whether the monitoring reports and related documents that meet CDM reforestation project conditions and tCER issuance requirements are in place;

Check whether the actual monitoring system and monitoring procedures are in line with the monitoring plan that is based on the registered methodology; check whether the record and storage of evaluation data are implemented according to the monitoring methodology.

In order to achieve the above objectives, verification is mainly based on the UNFCCC, and *Validation and Verification Manual 01.2*. It is divided into two steps: first, reviewing project documents, including project monitoring reports, PDD, project certification reports, afforestation monitoring cards, afforestation land contracts, construction contracts and related background materials; second, on-site verification, including physical check of project progress, monitoring system, setting and investigation of monitoring sample plots, confirmation of project boundaries, quality control of monitoring data, opinions and comments of project participants, and review of CER calculation.

During the certification process, a great deal of time was spent on field inspection of the afforestation sites. The project will be stratified according to tree species/forestation model and the year of afforestation. The project will be divided into 22 strata and 102 sample plots will be set up. The verification team proposed to check each stratum to achieve full certification. The verification content of fixed plots include: confirming the center pile of the plot with GPS, determining the four side piles of the plot with a compass instrument, measuring row spacing of the afforestation site, counting the number of live standing trees in the plot, and measuring the average height and DBH of five trees in the plot. While checking the investigation of the fixed plots, the positioning of project (small class) boundary along the route is also determined by GPS, and the neighboring villagers participating in the project are interviewed.

The core of certification is the consistency of project activities and design documents, including: whether the original land types, locations of actual afforestation, afforestation tree species, afforestation technical specifications, and tending measures are consistent with the design documents; whether the monitoring card is consistent with the technical standards required by the construction contract; whether the sample plot layout, sample plot survey steps, survey equipment, survey quality control procedures are consistent with the project monitoring report and *The Monitoring Operational Procedures for "the Reforestation Project at the Pearl River Basin of Guangxi in China*"; whether the data and parameters cited in the project documents and the formula for calculating tCER have sufficient justification.

The requirements for certification are very detailed and almost harsh. For example, with respect to filed survey equipments, there are specific requirement on the types of equipment used, such as GPS and altimeter, and the method of correction and frequency of corrections. There will be standard demonstrations in fieldwork. In the process of tree survey, the position of DBH scale must be placed according to the operating procedures. In field verification, it is necessary to stop at any time and use GPS to vertify the project boundary, carbon boundary, or subdivision boundary.

If the project is implemented strictly in accordance with the design documents, even if some activities might change due to unforeseen factors, project results can be recognized by the verification team as long as sufficient evidence can be provided for justification or explanation.

5 Project Benefit

The project has been implemented for 6 years, and its benefits are gradually emerging, which are reflected in the following aspects:

(1) Develop the world's first afforestation/reforestation methodology approved by the CDM-EB

During project preparation, the expert team conducted a comprehensive and detailed baseline investigation through collection and in-depth study of a large number of relevant data. The team comprehensively applied related theories and research results to develop the first afforestation/reforestation methodology approved by the CDM-EB, i.e., CDM Reforestation Methodology on Degraded Land (AR-AM0001). The methodology creatively solves technical difficulties such as baseline, additionality, non-permanence, and leakage of CDM reforestation projects on degraded land. It fills the gaps in this field, and becomes a model for follow-up methodology of CDM

A/R projects and is widely accepted; A large number of baselines and monitoring methods in the methodology are directly cited by other methodologies, making positive contribution to the development of international CDM A/R methodology. The approval of the methodology also injected a strong boost to global CDM A/R project which was at a low point. It has contributed actively to the development of global CDM A/R project and the role of forestry in combating climate change. It is a milestone in CDM A/R project. In June 2007, an expert committee from the Department of Science and Technology of the State Forestry and Grassland Administration evaluated the methodology and identified it as highly innovative and operational. This achievement won the 2nd Liang Xi Forestry Science and Technology Award of 2007.

(2) Design the first CDM A/R project approved by the CDM-EB in the world

It is the first time to prose project methodology under CDM A/R rules. The methodology includes the boundary investigation method, baseline stratification method, land eligibility and project additionality justification method, as well as baseline net greenhouse gas removals by sinks, leakage, carbon stock change, and greenhouse gas emissions accounting method. Project stratification method, sample plot sampling and quantity determination method, boundary monitoring methods, carbon stock change monitoring method, greenhouse gas emissions and leakage monitoring method, quality assurance, and quality control procedures were also proposed to fill the gaps in the field. The pioneering and major innovations represent the highest level of China's forestry industry in this field. In November 2006, the project was officially approved by the CDM-EB and became the first CDM A/R project in the world to be officially approved for registration, providing a model for designing similar projects at home and abroad. The design won the Silver Award of 2010 National Excellent Engineering Design.

(3) Train expert teams with integrated capacities

The CDM A/R project preparation process is complex and the review rules are strict. At the same time, there are technical problems such as methodological development, additionality justification, preparation of design documents, and carbon measurement and monitoring. Therefore, it is necessary to organize an expert team to develop methodologies and carry out project design before the project can be possibly registered in the CDM-EB. In the initial stage of preparation, the project had organized a team of domestic and foreign experts. The project's baseline investigation and afforestation design mainly relied on local experts to complete key technical problems, such as methodological development, additionality argumentation, and project review and registration. Problems such as changes in tree species and afforestation technology encountered in the implementation process are resolved by domestic and foreign experts. Through concerted efforts, the expertise of domestic and foreign experts is given full play. They complemented each other and overcame numerous difficulties to ensure successful registration and implementation of the project. Meanwhile, by participating in domestic and international training and cooperation with foreign experts, experts at different levels are familiar with and master international regulations, review and approval procedures, design methods, carbon measurement, and monitoring methods for the carbon sink and afforestation project. The combination of project practice with theories has created a team of experts who understand international rules and master domestic policies and are familiar with the situation in the project area, which effectively enhances project implementation.

(4) Establish an effective project management model

The implementation period of CDM A/R projects is generally more than 20 years. Generally speaking, large-scale forestry companies are reluctant to invest in such projects with long cycle and obscure short term economic benefits. If a local afforestation company (or forest farm) independently implements such large-scale and complex projects, it can't sustain with its own economic strength and technical capacity. Therefore, it is necessary to determine an effective and sustainable project management model. The project adopts a combination of two local afforestation companies with two forestry farms, and contract one company with stronger economic strength and technical capacity to work as the project participant from Chinese side. The company filed a project permit application to the national climate change authority and obtained approval. The companies (or forest farms) are independent in terms of project implementation and operation management. This is not only in line with the requirements of the *Operational Measures for CDM Project Management*, but also easy to operate and accepted by companies. It is an effective management model that can be promoted in similar projects in China.

(5) Play a sound demonstrative role

As the project is the world's first successful project registered in the CDM-EB, it attracted numerous domestic and foreign experts, scholars, and news media. Since the implementation, foreign experts, scholars, journalists, and NGOs from the United States, Britain, Germany, Austria, Australia, Canada, Japan, and Vietnam came for visit. Domestic authorities and counterparts from Jiangxi, Sichuan, Yunnan, Guizhou, Heilongjiang, Inner Mongolia, Qinghai, and other provinces also came for filed study and discussed and conducted academic research on key issues such as forest carbon project design, application, implementation and carbon sink trade. Through the visits, inspections, seminars, TV reports, newspapers, and the Internet by domestic and foreign experts, the project played a demonstrative role globally, which raises the awareness and promotes the development of forestry carbon sink projects. At the same time, forestry technicians, government officials, and the general public at all levels have greatly raised their awareness to forestry carbon sink projects, energy conservation, and emission reduction. The implementation of the project also effectively promoted the development of China's forestry carbon sink projects, which is highly relevant to the development of a green and low-carbon economy and building a resource-saving and environment-friendly society. The World Bank's BIOCF and the Carbon Sink Management Office of State Forestry and Grassland Administration have used this project as a pilot for CDM A/R projects and also as research site for related technologies, methods, and carbon trading mechanisms of forestry carbon sink project. Number studies have been carried out in terms of technical methods and carbon trading mechanism.

(6) Effectively promote the implementation of forestry carbon sink projects in China and the world at large

While effectively absorbing and sequestrating carbon dioxide, the project explores carbon trading mechanism, demonstrates and summarizes the technologies and methods related to forestry carbon sinks. These technologies and methods have been compiled into the *CDM A/R Project Practical Guide*, *Operational Procedures for Monitoring CDM A/R Project* and *Research Reports on Implementation Experiences, Obstacles, and Improvement Approaches of "the Reforestation Project at the Pearl River Basin of Guangxi in China*". These documents have been submitted to the World Bank's BIOCF and the CDM-EB to provide scientific basis and practical experience for formulating and improving implementation rules, policies, and standards of international CDM A/R project, which is conducive to improving the special role and importance of forestry in mitigating climate change; it has also provided practical experience and lessons for future CDM A/R project, which has saved costs, improved effectiveness, and avoid detours of later projects; it contributes greatly to implementing forestry carbon sink projects in China and the world at large, thus it has significant international relevance.

(7) Greatly promote the foreign exchange and cooperation of Guangxi's forestry sector

The project introduced advanced forestry technology and project management experience in China and abroad, which promoted the integration of Guangxi Forestry and the world. It provided examples, experience, and data for drafting CDM A/R project rules and policies, developing carbon trading market and promoting sustainable development of local economy. It is a concrete measure for China to implement the Kyoto Protocol and is conducive to enhancing China's international status and the influence of Guangxi. It has laid down solid foundation and created favorable condition for Guangxi to participate in international forestry exchanges and cooperation in a wider range in more fields and at a higher level, particularly for introducing more international investment.

(8) Improve the awareness to ecologic and environmental protection

Prior to project implementation, not only the rural households in remote mountainous areas but also the people in counties and townships knew little about the grim fact of global climate change, let alone how to deal with it. Through continuous publicity and training during the preparation phase and implementation process, the positive role of forest carbon project in combating climate change has gradually been recognized by people, and in this process, people transformed such efforts into their own practical actions, and even have it elevated to the extent of "low-carbon" self-discipline.

(9) Achieve significant environmental benefits

The land used for afforestation is degraded land at remote areas. In the past few decades, these lands have few weeds and shrubs, with no value for farming. Local communities, farmers, forestry companies or forest farms are unwilling to engage since they can't see the potential. Therefore, these lands are deserted for long time, causing great waste and severe soil erosion. The implementation of the project leads to greening of 3,008 ha of deserted and degraded land. The barren hills are gradually covered by lush forests, effectively reducing soil erosion. With forest growth and increase of forest biomass, the amount of carbon dioxide absorbed also increases. Eco-service value is gradually emerging in terms of forest ecosystems, water resources conservation, soil conservation, carbon sequestration and oxygen release, nutrients accumulation, atmosphere purification, biodiversity conservation, and forest recreation. It is estimated that the value of newly added forest achieves RMB 850 million of annual eco-service function. Moreover, benefits are gradually showing through project implementation, which causes local forest farms (or forestry companies), communities and farmers to pay attention to and use similar degraded lands, so that more deserted and degraded lands can be fully utilized to contributed to promoting local eco-environment.

(10) The project has obvious social benefits

Through project implementation, the skills and income of forestry products in the communities and farmers have been significantly improved. During the implementation, provincial and county forestry authorities organized several types of training courses. The training takes various forms such as convening villagers meetings, interviews with farmers, on-site guidance for afforestation, and distribution of materials. The training covers forestry and knowledge related to global climate change, CDM A/R project implementation rules and international and domestic policies, afforestation species and management mode selection, afforestation technology and forest management measures. At the same time, the first phase of the project benefits more than 5,000 farmers in 10 villages (towns) and 27 villages in the project area (including 1,793 minority beneficiaries); the second phase benefits more than 19,180 farmers in 28 villages (forest farms) and 74 villages (sub farms) in the project area (of which 86% are minority beneficiaries). The benefits include: since seedlings, afforestation and tending all require investment of a large amount of labor, it provides employment opportunities for local surplus labor force, and increase farmer's labor income; whether it is associated with company (forest farms) or leased land to company (forest farm) for afforestation, community (farmers) can get rental income or expected forest products revenue;

Obtain carbon sales revenue. At present, although there is not much income from carbon, for farmers living in remote and impoverished mountainous areas, they are greatly motivated to participating in the carbon project, since that they can get an extra income from the deserted and almost forgotten land. Originally they all thought that "selling air to make money" was only a dream. Now it has become a reality, which is a bid surprise for them. Therefore, they have increased their initiative and enthusiasm in participating in projects and managing project forests.

6 Experience and Lessons

The "reforestation project at the Pearl River Basin of Guangxi in China" is an unprecedented project with wide attention. From preparation to implementation, it is highly innovative and challenging.

6.1 Experience

(1) The guidance and support of government departments are crucial

The CDM A/R project is a new type of project. It can be said that before 2004, neither the Department of Forestry of Guangxi Zhuang Autonomous Region (management authority), the Guangxi Forestry Inventory & Planning Institute (technical agency), nor participating companies, forest farms, and farmers (project implementing entities) knows anything about forestry carbon projects. It is hard to imagine that such a challenging project can be done by these companies, forest farms, and farmers. Therefore, the leadership and full support of government is of utmost importance. First, at national level, the national climate change authority and the State Forestry and Grassland Administration provided lots of suggestions and support in terms of participant qualifications, review and approval procedures, CER price and quantity, so that the project is securely and successfully approved. Second, at organizational level, Guangxi set up special project management offices in the forestry departments at district and county level responsible for project reporting, preparation and implementation guidance. At technical level, district and county project offices are responsible for organizing and contact Guangxi Forestry Inventory & Planning Institute and Guangxi University to carry out specific work such as baseline investigation, afforestation design technical training and monitoring; in terms of funding, in addition to appropriate ecological subsidies for afforestation, forestry departments also organize and fund such costly events as training, baseline investigation and fixed samples plot monitoring. Even for some detailed and specific issues, such as disputes arising from joint ventures, inspections of afforestation quality, and payment and issuance of carbon remittances, government coordination and assistance is indispensable.

(2) Professional technical support is the essential of project's success

The preparing and implementing CDM A/R projects face some difficulties. First, people have to be familiar with relevant international and domestic rules; second, they have to understand the characteristics of forestry carbon projects, and grasp forest management knowledge and participatory community survey methods, which is difficult with respect to the technologies involved. From a large number of fieldworks such as woodland selection, baseline investigation, afforestation design, and sample plot monitoring, to a series of desk working such as project concept documents, carbon financing documents and PDD, qualified and experienced professionals are required. Fortunately, the project received technical guidance from the World Bank, BIOCF, the State Forestry and Grassland Administration, the Chinese Academy of Forestry, and the nature conservation experts. Experts and technicians of the Guangxi Forestry Inventory & Planning Institute played a pivotal role in this, and professionally completed a series of work in a professional manner, such as baseline investigation, afforestation design, archives creation, sample plot layout, and carbon reserve monitoring.

The project design mainly relied on the strength of domestic experts. International expert also provided recommendations in key technical aspects such as methodology development, project certification, and registration. Experts were given full play to their capacity and complement each other in the process. This proves to be critical in ensuring successful registration and smooth implementation of the project.

(3) Reasonable project organization and implementation form and income distribution method are key to ensuring the continuous project implementation

The right to land contractual management of the project is owned by villagers' collective group and farmers, involving thousands of households. In order to make the project meet the design requirements, and be acceptable to project afforestation entity and land manager, the project expert team, through repeated communication and discussion with local authorities, project afforestation entity and land manager, finally established two forms of project organization and implementation, namely villagers' collective group (farmers) and forest farms (company) cooperate by share, and villager groups. At the same time, a participatory method was adopted to establish benefit distribution of timber revenue and CER trading in the wake of afforestation. After gaining the approval from various stakeholders, it was determined in the form of a contract. Both organizational forms and benefit distribution methods are in line with China's current conditions and forest development, and have been applied in similar domestic projects. Through project implementation, it is shown that, when determining the implementation and benefit distribution method, the most important thing is to streamline the relationship between project participants and implementing entities and landowners, especially the benefit distribution, only by fully securing the legitimate interests of local communities (the farmers participating in a project in particular) can the long-term and sustainable project implementation be ensured.

(4) Effective management model is an important guarantee for continuous and orderly project development

As the first CDM reforestation project that is successfully registered in the UN CDM-EB and the carbon sink project pilot project jointly established by the Department of Forestry of Guangxi Zhuang Autonomous Region and the BIOCF, it has to stand the verification of designated operational entity and the test of time. Meanwhile, the project covers a wide range of areas with long implementation period, and its implementing entities are mostly small companies, forest farms, and farmers. Therefore, it is imperative to establish a long-term management model. To this end, the project established a complete three-tier management model: the first tier is the project office of Forestry Bureau of the Autonomous Region, which is responsible for organizing project application, liaising with the BIOCF and related national management agencies, organizing project training, checking the progress and quality and organizing expert team to complete project monitoring. The second tier is the county project office and the expert team. The county project office is responsible for formulating the project's annual implementation plan, carrying out technical training, organizing technical personnel to cooperate with the expert group in conducting project monitoring, preparing and reporting the required materials, and establishing and managing county-level (including farmer households) project archives; the expert team is made up of experts from Chinese Academy of Forestry, Guangxi Forestry Inventory & Planning Institute, Guangxi Zhuang Autonomous Region Forestry Research Institute, who are responsible for formulating project implementation guidelines, inspection and acceptance methods, project monitoring procedures, project training materials, and forest growth standard. The team assists district/county project offices in solving major technical problems during project implementation, conducts quality supervision on project activities of the afforestation entities, undertakes project monitoring and analysis of project monitoring results, and prepares monitoring report. The third tier is the project implementing entity. Under the organization and leadership of district and county project offices, the afforestation and management are carried out. Project activities are recorded truthfully according to the project operating guide and monitoring manual. The small-class record card and leakage monitoring log are filled in and reported to county project office on time. Project office and the original records are well kept.

The scale of the "the reforestation project at the Pearl River Basin of Guangxi in China" is small, but the three-tier management mode is indispensable. Only by keeping the stability of project staff and maintaining the long-term existence of management organization can the project be sustainable.

(5) Comprehensive and systematic technical training is a powerful guarantee for improving project execution

The development of methodology, the demonstration of additionality, the preparation of design documents, the carbon measurement and monitoring are the main technical difficulties in the current CDM A/R project development. During project preparation and implementation, with respect to project background, project characteristics, technologies and management issues, international and domestic experts were invited to conduct comprehensive and systematic trainings for participant from different levels including authorities of the Autonomous Region, design agencies, local authorities, the implementing entities, and local communities. By combining with project practice, the stakeholders at all levels of the project have a better understanding of the specific requirements and rules for carbon sinks afforestation projects, as well as review and approval procedures, design methods and carbon reserve measurement and monitoring methods. This has effectively enhanced the execution of project participants and ensured the smooth development and implementation of the project.

6.2 Lesson Learned

The experience gained from the project is of positive significance and the lessons learned are also a valuable asset. To sum up, this project has lessons to be learned in the following areas:

(1) The selection of afforestation sites must be careful

Land eligibility is a prerequisite for the successful registration of the project as a CDM project, and it is also the basis for project development. If the selected land fails, there won't be any project to talk about. Regarding the degraded and deserted land over 50 years or since December 31, 1989, it is a very clear concept. However, if this is the only criteria without considering other factors, the project will be put in a very embarrassing situation. As far as this project is concerned, in order to meet land eligibility, it seems that the more remote, more deserted, and more barren the land is, the better it is for the project. The result is that the land is agreed upon, the project is approved, and project registration is successful, but when it comes to the implementation, the difficulty comes out. The remoteness of the project site caused a sharp increase in labor input. The poor land condition makes trees grow slowly, and the combined effect is that the input and output ratios are disproportionate. The implementation entities such as companies, forest farms, and farmers are all expecting certain economic returns from the project. In reality, some places are facing severe condition, which affects the enthusiasm of forestry owners and will inevitably affect the sustainability of the project. Therefore, the lesson to be learned is that people should not go to extremes in land selection. Land eligibility is important in that without such land no CDM A/R project is achievable. However, based on that, it is also necessary to consider the productivity of the selected land and its traffic conditions should be with acceptable range.

(2) The ownership of afforestation land must be clear and stable

Although the land for the project was negotiated with the farmers through participatory design and a forestry agreement was signed, the reform of collective forestry right system in recent years has changed the ownership of forest land. At the time of project design, part of the contracted forest land management right was collectively owned, and the contract was signed between afforestation entity and the villagers. After forest ownership reform, these leased or jointly used lands were contracted to farmers. Some of the farmers ignored the legal effect of the contract and took advantage to raise unreasonable demand. They even tore up the contract and withdrew from the project, which caused certain part of planned sites unable to be completed.

(3) The selection and configuration of afforestation tree species must be scientific and reasonable

The CDM A/R project requires that afforestation sites be without any trees since December 31, 1989. The project site selected according to this standard is part of a barren mountain located at a high altitude, with poor traffic, poor site conditions, steep slopes, and high wind. At present, there are few afforestation trials and demonstration studies conducted on these difficult sites. People have little practical experience in what tree species should be planted and what kind of afforestation technology should be adopted. The project has no precedent to follow in terms of selecting afforestation tree species and afforestation technology design. They can only blindly refer to the experience of afforestation in other similar regions in a theoretical approach. Due to unreasonable selection of afforestation tree species and poor site conditions, some afforestation failed or resulted in poor tree growth, which greatly affected the enthusiasm and confidence of afforestation entities in implementing the project. Meanwhile, CDM A/R projects have very strict requirements on afforestation sites, land boundary, afforestation tree species and allocation, and forestry technology. Once the project design is approved by an independent operating entity, it must be strictly developed in accordance with the design, and it is difficult to change. Therefore, the ideology of conventional afforestation must be broke down, and tree species have to be selected and configured in a scientific, rational, and prudent way. Due to the lack of supervision of the afforestation entities and related departments during project implementation, plus the impact of land and natural disasters, the project could not be implemented according to the designed afforestation year and tree species can't be allocated accordingly, which caused great trouble for project monitoring and increased project risk at the same time.

(4) When selecting a project, it is necessary to understand the difficulty, cost, and risk of the project

First, the project preparation time is long and difficult. When developing CDM A/R projects, it is necessary to select or develop appropriate project baselines and monitoring methodologies. If the approved methodologies are not suitable, new methodologies must be developed and submitted to the CDM-EB for approval before they can be used for project design. The development of new methodologies not only requires lots of manpower, material, and financial resources but also has technical difficulties such as baselines, additionality, non-permanence and leakage, thus they may not be approved. This means that developing new methodologies will face greater risks. After completing the design documents, before submitting for registration to the CDM-EB, in addition to the approvals from domestic climate change authorities, an operational entity will be designated by the CDM-EB to carry out a full review of the project's conformity (including on-site inspections) in accordance with relevant international rules. The validation report needs to be publicized on relevant website, and all questions posed by relevant experts should be answered during the notification period. The whole process is very complicated and takes quite a long time. From project preparation to the approval of the CDM-EB for registration, it took nearly three years.

Second, the project has a long implementation cycle and high risk. At present, the implementation period of CDM A/R project is 20 or 30 years, which means that the project will be subject to the risk of natural disasters such as forest fire, pests and diseases, and drought. According to statistics, at the beginning of 2008, 100.6 ha of forest in the project was reforested due to the impact of snow and ice; in the autumn of 2008, some young trees were also affected by drought, and the survival rate was so low that they needed to be replanted; in the spring of 2011, the *eucalyptus*, which suffered freezing hazard in 2008, was devastated again by freezing disaster. This has greatly increased afforestation cost and brought heavy economic burden on forestry owners. In addition, most of the afforestation areas are located in remote mountainous areas and the transportation is inconvenient. This brings great difficulties to material transportation and afforestation, which in return, increases the cost of afforestation and the difficulty of implementation.

Third, forestry carbon market has small demand and the carbon price is low. The CDM forestry carbon has no market globally and the price is low. Generally speaking, the price is below US$5/ton. Since USD is continuously going down in the past 10 years. The exchange rate of USD to RMB has dropped from 1:8.09 (project signature) to the current 1:6.30. This price is disproportionate with respect to the implementation difficulty, project application and implementation, and the high cost of monitoring. In addition, social and environmental impacts impose more constraints on the implementation of projects, which makes the project less attractive to forestry owners and the implementation more difficult.

References

Compilation Committee of the Records of Cangwu County (1996) Record of Cangwu County
Guangxi Huanjiang Xinghuan Yinglin Co., Ltd (2011) Implementation report of the reforestation project at the Pearl River Basin of Guangxi in China
Guangxi Forestry Inventory & Planning Institute (2006) Project design instructions for the reforestation project at the Pearl River Basin of Guangxi in China
Guangxi Forestry Inventory & Planning Institute (2012) Monitoring report for the reforestation project at the Pearl River Basin of Guangxi in China
Compilation Committee of Huanjiang Maonan Autonomous County (2011) Record of the Huanjiang Maonan Autonomous County
NDRC, MOST, Ministry of Foreign Affairs, Ministry of Finance (2005) Operation and management methods for clean development mechanism project

Afforestation and Reforestation Project on the Degraded Land in Northwest Sichuan, China

Caifu Tang, Jian Ma and Biao Yang

Summary

The "Afforestation and Reforestation Project on the Degraded Land in Northwest Sichuan, China" is the first registered CDM A/R project that has received a gold certificate with the Climate, Community & Biodiversity (CCB) standard. It is also the second in China and the first in Sichuan to be successfully registered as a CDM A/R carbon project. The project is led by the Forestry Department of Sichuan Province, with the support from the State Forestry and Grassland Administration (SFGA), the Conservation International (CI), The Nature Conservancy (TNC), and the Beijing Shanshui Conservation Center, Funded by 3M, the project is launched in November 2004, with the Dadu River Afforestation Bureau of Sichuan Province (hereinafter "the Bureau") as the operating entity. In July 2007, the PDD was completed. In April 2008, the DOE field verification was passed and the project was approved by the Forestry Bureau and approved by the SFGA. In June 2009, the project passed DOE validation. In November of the same year, the project was endorsed by the NDRC and the CDM-EB registration under the UNFCCC.

The project is located in the southwest mountainous region of China, which is among one of the biodiversity hotspots. The project plans to implement 2,251.8 ha of plantation afforestation in 28 villages of 21 townships, covering 5 counties (Li County, Mao County, Beichuan County, Qingchuan County, and Pingwu County) in Sichuan Province. Upon completion, it is estimated that a reduction of 460,603 tons of

C. Tang (✉)
Sichuan Green Carbon Ltd., Chengdu, China
e-mail: 631592345@qq.com

J. Ma
Paradise Foundation, Beijing, China
e-mail: mj@pfi.org.cn

B. Yang
SEE Foundation, Beijing, China
e-mail: yangb315@163.com

Z. Lu et al. (eds.), *Forest Carbon Practices and Low Carbon Development in China*, https://doi.org/10.1007/978-981-13-7364-0_5

CO_2 equivalent can be achieved during the 20-year crediting period, with an average annual amount of 23,030 tons; 3,231 rural households (12,745 farmers in total, of which 42.2% are ethnic minorities) from 5 counties in the project area will benefit from employment, carbon sinks, wood, and forest by-products. The annual per capita income will be increased by 10.68% on the basis of 2006; the project is expected to create 1.105 million short-term job opportunities and 38 long-term jobs for local or neighboring farmers. It is expected that the following objectives will be achieved through project implementation and operation: restore forest vegetation with multiple benefits, enhance the capacity to absorb carbon dioxide from the atmosphere and mitigate climate change; improve water conservation of forest and improve regional ecological environment; facilitate the connectivity of neighboring forest ecosystems around the nature reserves and contribute to the conservation of biodiversity; increase the income of local villagers and alleviate the pressure on poverty.

The project starts in 2007. Currently, afforestation has been completed for the construction period, and has entered maintenance and management for the operation period. Meanwhile, the first verification period (2007–2012) of the project has been achieved and a trade cooperation agreement of about 50,000 tons of carbon dioxide emission reduction was signed.

1 Project Background

In 2004, with the funding from 3M, a Fortune 500 company, CI, the Beijing Shanshui Conservation Center and the SFGA worked together, under the goal of conserving biodiversity and improving local communities' livelihood, to launch the Forest Restoration project for Climate, Community & Biodiversity (FCCB project), with the purpose of restoring natural vegetation, exploring innovative conservation financing mechanism based on global CO_2 emission trade market The project employed with the Climate, Community & Biodiversity (CCB) standard to develop carbon projects. While mitigating global climate change, it conserves key areas with ecological conservation and works to improve the livelihood of the local community. The southwestern mountainous region of China is one of the 25 biodiversity hotspots in the world. It is known as the "water tower of significance", the "core area of biodiversity," and the "sensitive area of global change". The unique combination of complex terrain and favorable moisture has resulted in an extremely rich biodiversity that boasts a large number of unique plants and animal species, which probably hosts the richest plant species in the temperate regions of the world. Of the more than 12,000 higher plant species found in the area, 29% are exclusive to this hotspot area. More than 300 mammals and 686 bird species were recorded. There are also a large number of locally exclusive animals and rare and endangered species in this hotspot area, including giant pandas, red pandas, golden monkeys, snow leopards, takins, Sichuan sika deer, cormorants, white-lipped deer, and 27 species of pheasant. The hot spot area accounts for about 10% of China's geographical area, but it is home to about 50% of the country's birds and mammals and more than 30% of higher plants. Of the 87 endangered terrestrial mammal species in China, 36 are in this region.

Cultural diversity flourishes together with biodiversity. Seventeen of the 55 ethnic minorities in China live in the area, including Bai, Dulong, Lisu, Naxi, Pumi, Nu, Yi, and Tibetan. Each ethnic group has its unique understanding of the natural resources in this unique environment. After several generations, these understandings have formed a reservoir of the knowledge base. Several of Asia's largest rivers also run through the area, including the Brahmaputra River (YalungTsangpo River), the Ayeyarwady River, the Mekong River (the Lancang River), the Salween River (Nu River), and the Yangtze River. These rivers add up to affect the lives of more than 500 million people living on about 3 million square kilometers downstream.

The project area is mountains area with the average altitude of 2,000–3,500 m. Most of the forests at the project site were degraded in 1950–1980 and the soil erosion was serious. The project area is rich in biodiversity, with 2 national nature reserves and 7 provincial nature reserves. The reserve areas recorded more than 3,000 species of higher plants, 85 species of mammals, 210 species of birds, 18 species of amphibians, 19 species of reptiles, 12 species of fish, and 287 species of insects. Among them, 7 species are protected plants of national category I, 6 species are protection plants of national category II; forest musk deer, alpine musk deer, giant pandas, golden snub-nosed monkeys, takins, snow leopards, and clouded leopards are all listed as protected animals of national category I; green-tailed rainbows pheasants, chestnut-throated partridge, Chinese grouse, golden eagles, and black-necked cranes are among the protected birds of national category I. Local residents in these remote mountainous areas (mostly ethnic minorities such as Tibetan, Yi, and Hui) are living below the poverty line.

This is a forest carbon project funded by enterprises and jointly developed by governments and NGOs. It is the first time to introduce the concept of FCCB in the afforestation project and CCB standard is used to develop and design projects. The successful development and practice of the project have a very important demonstrative effect on the development of China's forest carbon project.[1]

2 Project Milestones

On November 8, 2004, FCCB project was launched in Sichuan Province.

On November 22, 2004, the Forestry Department of Sichuan Province held a workshop on the selection of carbon projects to clarify the basic principles, methods, and indicators and criteria of the project.

---

[1] FCCB project upholds the purpose of protecting biodiversity and improving people's livelihood. Located in one of the hotspots of global biodiversity, namely the southwest mountainous regions of China, it is a forest, Climate, Community & Biodiversity project that aims at restoring natural vegetation, mitigating climate change, promote community development, protecting biodiversity, and exploring multiservice function of forest ecological system.

From December 2004 to January 2005, the Forestry Department of Sichuan Province organized experts to conduct a site selection survey.

From January 22 to 23, 2005, CI held a seminar on FCCB project in Chengdu. A total of 24 officials and experts from the SFGA forest carbon office, the Forestry Department of Sichuan Province and other related organizations participated in the event to discuss project investigation form, project flow, and work list.

On February 24, 2005, 3M, a world-renowned Fortune 500 company, announced in Beijing that the 3M donated US$3 million and would work with CI to implement an FCCB project in southwest China. Leaders of national authorities, officials of Sichuan Province and Yunnan Province, as well as project representatives of CI and other guests attended the press release to witness the official launch of the project.

From March 7 to 10, 2005, the officials and experts of the SFGA carbon sink office, the CI, and the Forestry Department of Sichuan Province visited the Wang Lang and Tangjiahe National Nature Reserves, as well as Pingwu County and Qingchuan County (optional sites). A discussion was made at the Forestry Department of Sichuan Province with respect to the project and site selection.

From April 21 to 22, 2005, the Forestry Department of Sichuan Province organized a training course of baseline data collection for potential forest carbon projects in counties, with 39 participants from 33 counties (cities, districts) and related units of the Forestry Department of Sichuan Province.

From June 8 to10, 2005, the Forestry Department of Sichuan Province organized a training course of village-level basic information collection for the forest carbon project. A total of 20 people from the project county, including key technical staff and Sichuan Forestry Inventory and Planning Institute participated in the event.

In September 2005, the first draft of the *Forest carbon Investigation and Design Guidance* (trial version) of Sichuan Province was completed. It was verified, modified, and improved through testing. In December, the document was renamed as the *Investigation and Design Guidance on CDM Afforestation and Reforestation Carbon Project (trial version)*, which is used to guide the investigation and design of forest carbon project in Sichuan.

From October to November 2005, the investigation and planning team organized by the Sichuan Forestry Investigation and Planning Team started a preliminary background investigation and project design.

From October 23 to 27, 2005, a seminar on FCCB standard and carbon monitoring and monitoring methodology were held in Longquanyi, Sichuan Province. 61 experts and project county technicians from Sichuan Province and Yunnan Province attended the event. Foreign experts were invited to give lectures and they also held discussions with senior officials of the Forestry Department of Sichuan Province and related departments.

In December 2005, the Sichuan Forestry Inventory and Planning Institute completed the development of forest carbon projects database. From March 27 to 30, 2006, a training course on application software for forest carbons was held in Pingwu County. 25 engineering and technical personnel from the project counties and experts from the Sichuan Forestry Inventory and Planning Institute participated in the event.

From July 5 to 6, 2006, a training seminar on developing a methodology for forest carbon project was held in Pingwu. Forty-seven people from 12 organizations participated in the event.

From July 7 to 11, 2006, 10 experts and personnel engaged in the implementation of forest carbon project in the province were invited by the forest carbon office of Forestry Department of Sichuan Province and CI to conduct field study of potential project sites in Pingwu County, Qingchuan County, Beichuan County, Mao County, and Li County.

From August 2 to 4, 2006, the "Training Course on CCB Standards and Community Assessment Methodology" was held in Chengdu. 23 participants from 9 entities participated in the event.

From August 9 to 13, 2006, the forest carbon office of Forestry Department of Sichuan Province and CI held the "Training on Carbon Baseline Investigation and Practice of FCCB Projects in Sichuan" in Li County. 26 people from 11 entities participated in the event. This practical training combined intensive lectures with field practices.

In September 2006, initial project investigation, carbon baseline investigation, community surveys, and field work of afforestation operations were completed. In October, data and graphics were completed, as well as the first draft of relevant reports.

In October 2006, the Sichuan Academy of Forestry completed a report on the biomass of main shrub species in northwestern Sichuan. The report put forward biomass models of four main shrub species, namely coriaria nepalensis, emei rose, wild pepper, and berberis kawakamii hayata, which laid down the foundation for estimating baseline investigation of the shrub.

From October 26 to 27, 2006, the forest carbon office of Forestry Department of Sichuan Province organized project personnel to participate in the first Carbon Expo Asia held in Beijing to publicize and display FCCB projects and formally launched the Panda Carbon—"Afforestation and Reforestation Project on the Degraded Land in Northwest Sichuan, China".

In May 2007, Sichuan completed the review of the materials required for PDD. The Dadu River Afforestation Bureau signed a contract with the community.

In July 2007, the PDD was completed.

In September 2007, the revision of afforestation design in 5 project counties was completed.

In mid-April 2008, TüV-SüD (Germany), working as a DOE, visited some of the afforestation plots and communities in Li County, Mao County, Qingchuan County, and Pingwu County to conduct on-site verification of project conformity.

On September 10, 2008, the project passed the review of the 52nd meeting of the National CDM Project Review Board. On November 11, the NDRC approved the project as a CDM project, and authorized the Dadu River Afforestation Bureau to implement the project as the implementing entity of the Chinese side.

On November 16, 2009, the project was registered with the CDM-EB under the UNFCCC.

On November 26, 2009, the Dadu River Afforestation Bureau, Hong Kong Carbon Care Asia and the Beijing Shanshui Conservation Center held a signing ceremony of forest carbon trade in Chengdu, and successfully signed on the first verification period (2007–2012) of the project, which covers approximately 50,000 tons of CO_2 emission reduction trade.

On January 28, 2010, the project's promotional event was held in Hong Kong SAR. More than 30 people from 30 enterprises and institutions attended the event, including the Hong Kong Jockey Club, Hong Kong Airport Authority, Asia World Expo, Town Gas, New World Development Company Limited, Jebsen Group, and Johnson Electric.

On June 8, 2010, the Forestry Department of Sichuan Province held a follow-up meeting on the forest carbon project in northwestern Sichuan. The meeting summarized project progress, discusses new afforestation, replanting, tending and management work, and carbon publicity and sale; a project management coordination committee was set up to clarify the input and the allocation of carbon revenue among various stakeholders on April 11, 2011, the Forestry Department of Sichuan Province held a discussion meeting on the forest carbon project in northwest Sichuan in Leshan. The meeting summarized the project implementation and arranged project follow-up.

3 Project Content

3.1 Project Area

The project is located in the northwest of Sichuan Province, including 5 counties, i.e., Li County, Mao County, Beichuan County, Pingwu County, and Qingchuan County. It also involves 28 villages in 21 townships, including 5 villages in 2 townships of Li County, 1 village in Mao County, 13 villages in 10 townships of Qingchuan County, 4 villages in 4 townships of Pingwu County, and 5 villages in 4 townships of Beichuan County. The project plans to establish 2,251.8 ha of plantation forest with multiple benefits on degraded land, and it is expected to generate 460,000 tons of tCER of CO_2 equivalent over the 20-year crediting period, with an annual average of 26,000 tons of CO_2 equivalent.

3.2 Project Target

Through project implementation, the forest vegetation in the mountainous areas of southwest China can be restored and the FCCB can be realized. These are reflected in:

- Increasing the capacity of forests to absorb CO_2 and mitigate climate change;
- Improving the connectivity of forest ecosystems around the nature reserves, conserving and establishing biodiversity corridors;

- Improving the water and soil conservation in the project area and improving the regional ecological environment;
- Increasing villagers' income in the community and elevating the income and living conditions of the villagers;
- Exploring a new mechanism of Payment for Ecosystem Services and providing demonstrations for carrying out multi-benefits forest restoration and carbon projects.

3.3 Implementing Entity

The owner of this project is the Dadu River Afforestation Bureau. The bureau was formerly the Dadu River Timber and Water Transport Bureau. Established in 1956, the Bureau is mainly responsible for timber transport and sales in western Sichuan. It contributed to the country's economic development. In 1993, in accordance with the Central Government's requirements for establishing a modern corporate system in state-owned enterprises, the Dadu River (Group) Co., Ltd. was established, with the Dadu River Timber and Water Transport Bureau as the main entity. On September 1, 1998, after implementing the natural forest conservation program, the Bureau stopped all timber transport, sales and related operations, and shifted its focus to the natural forest conservation. On May 11, 2001, with the approval of the Forestry Department of Sichuan Province, the "Dadu River Afforestation Bureau of Sichuan Province" was registered, signifying that the Dadu River Timber and Water Transport Bureau had a major shift from drifting in the river to planting trees. By the end of 2007, there were 2,311 full-time employees, 3,912 retired employees, and 970 additional workers with pension and subsidies. At present, the Bureau has 1,983 on-duty employees, 4,036 retired employees, and 1,101 additional workers with pension and subsidies. There are 5 senior engineers, 30 engineers, 33 assistant engineers, 45 technicians, 9 accountants, 15 assistant accountants, and 16 account clerks.

3.4 Operation Model

The project is funded by 3M, with CI, the Beijing Shanshui Conservation Center and TNC providing project development and technical support. This includes personnel training, project feasibility study, PDD, national review, DOE verification, CDM-EB registration, CCB certification, carbon trading, and carbon monitoring. The operating entity—the Dadu River Afforestation Bureau, signed a land use contract with landowners and county forestry bureaus. During the implementation stage, the "Project Management and Coordination Committee" was set up, with senior officials and personnel of the Forestry Department of Sichuan Province, the Beijing Shanshui Conservation Center, the Dadu River Afforestation Bureau, and 5-county forestry bureaus providing project coordination and management services.

Farmers and communities in the project area provide land to ensure that there is no dispute over forest land rights and do not undertake afforestation investment; community villagers can obtain certain income by participating in projects such as soil preparation and planting. The project owners raise funding for afforestation, the main source of which is to apply national fund for seedlings, and the companies will cover the operating cost by themselves.

All project counties are responsible for forest management, such as tending, replanting, management and protection, and forest supervision. They are responsible for all plantation costs, and are responsible for the cost of monitoring, verification, sales, and coordination that is related to carbon within each county. The Dadu River Afforestation Bureau is responsible for the sale of carbon, small-scale sales promotion and publicity, monitoring fees, and DOE verification fees, In terms of benefit allocation: 30% of carbon income belongs to forest owners or farmers, who also own all forest and forest by-products; 15% is public fund, which is used by the Management Coordination Committee of the Project for coordination, management, training, learning, field study, and other activities; 10% is for technical support and carbon sales coordination costs; for the remaining 45%, 25% belong to the Dadu River Afforestation Bureau, 20% to the county forestry bureaus. After the end of the project cycle, the land and forest are returned to land owners, and the project owners no longer participate in sharing the benefit.

3.5 Project Duration

The crediting period of the project is 20 years, from January 2007 to January 2027, and can be renewed twice.

3.6 Land Eligibility

In accordance with the definition of forests defined by the Chinese government, "Land eligibility procedures for afforestation/reforestation project activities" of the United Nations CDM-EB is used to justify land eligibility:

Field study indicates that the selected land is degraded, low-yield wasteland, or illegal grazing land. The current vegetation cover is mainly herb and sporadic shrubs, which does not meet the forest standards defined by the Chinese government; the project afforestation site is not the non-forest or plantation forest land that can meet the forest standard defined by the Chinese government without people's intervention; at least since 1989, the project land is not a temporarily no-forest land due to harvesting or indirect natural causes such as fire and pests; grazing, firewood collection and lack of source seeds affect the renewal of natural vegetation and cannot meet the standards defined by the Chinese government.[2]

The latest (year 2000) land use/land coverage mapping land is used to prove that the selected project site has no-forest coverage.

Through interviews with local farmers, the history of land use/coverage and important activities affecting land use/coverage is gathered, which states that the selected project site has been forestless since at least 1989, and the project site in Li County has been a no-forest land for the past century.

The land use/coverage mapping before 1990 is also used to prove that the project land was a no-forest land before 1990.

3.7 *Baseline Scenario*

According to the actual situation of the project, the revised methodology AR-AM0003/03 was selected to demonstrate the most reasonable baseline scenario:

(1) Determine project boundary

The land use/land coverage mapping and the farmers' interview are used to verify land eligibility. Then, GPS is used to determine the coordinates of the location point (latitude, longitude) on each plot.

(2) Analysis of land use

From one century ago to the 1980s, the land within the project boundary was historically a forest land. However, large-scale deforestation took place. At present, these lands are degraded, low-yield wastelands, or grazing lands. Vegetation cover is dominated by herbs, shrubs, and sporadic trees. According to the local land use plan, the land within the project boundary are all forestry lands, thus grazing within the project site is illegal. Currently, the coverage of scattered trees in the project area is less than 20, which cannot meet the standards defined by the Chinese government for forests. Due to the disturbances such as grazing, fires, firewood collection, and soil erosion, natural regeneration can't enable the project site to achieve 20% in the future as specified for defining trees.

Through interviews with local farmers, it is learned that more and more serious illegal grazing, fire, firewood collection, and soil erosion have led to a continuous decline in the coverage of trees and non-tree vegetation over the past few decades.

For many years, the local government has been insufficient in funding for afforestation/reforestation, and the money obtained is mainly used for afforestation on land with good conditions and within close distances. The land selected for the project is relatively remote and timber is not subject to full market competitiveness. The introduction of CDM A/R projects will not affect the existing afforestation/reforestation projects.

[2]http://cdm.unfccc.int/Reference/Procedures/methAR_proc03.pdf.

Field surveys, farmers' interviews, and socioeconomic analysis indicate that a possible land use method is to maintain current land use status (maintaining the state of wasteland or continuing with illegal grazing) or to conduct afforestation activities, since the use of these wastelands is restricted by the government, for example, farming and animal husbandry activities are prohibited. If it is restored to forest land, investment analysis shows that when there is no carbon income, the project's FIRR is only 5.36%, which is far below the 8% of the required rate of return (RRR) for government forestry project.

(3) Division of Baseline Carbon strata

First, field study is conducted on each site to investigate the existing vegetation, soil conditions, and erosion; second, interviews were conducted with local farmers to understand the history of land use/coverage, important activities affecting land use/coverage, and current human activities (grazing, firewood collection, etc.); Finally, based on the information obtained from field study and interviews, it is confirmed that the existing vegetation and grazing activities are the most important factors affecting current status and baseline. Based on that, the project is divided into 14 carbon baseline strata (Table 1).

(4) Determination of baseline carbon stock

The random distribution of geographic information system was used to distribute sample plots directly on the map, and plots were randomly sampled for investigation.

The investigation of herb biomass was done in smaller sample plots, with the size aboveground at 1 m × 1 m and belowground at 50 cm × 50 cm; for the shrub biomass, above-ground and below-ground investigation was sampled at 2 m × 2 m and at 1 m × 1 m; for the four major large shrubs (coriaria nepalensis, emei rose, berberis kawakamii hayata, and wild pepper), the sample size is 5 m × 5 m. Scattered wood is measured within a radius of 20 m.

In addition to large shrubs, the direct biomass of above-ground and below-ground herbaceous plants and shrubs is measured by harvesting and weighing fresh biomass within the project site.

The biomass of the four main large shrubs (coriaria nepalensis, emei rose, berberis kawakamii hayata, and wild pepper) can be obtained by allometric growth equations. The specific formula is as follows:

Coriaria nepalensis: $SB = 129.1004\ (DB^2 \cdot SH)^{0.9011}$
Flat sting rose: $SB = 4.7748\ (DB^2 \cdot SH)^2 + 101.72 \cdot DB^2 \cdot SH + 14.628$;
Berberis kawakamii hayata: $SB = 336.447 \cdot DB^{1.0463} \cdot CD^{1.0348}$;
Wild pepper: $SB = 93.9375 \cdot (DB \cdot SH)^{1.5869}$;

Among these, SB, shrub biomass, kg; DB, shrub diameter, cm; SH, shrub height, meter; CD, shrub crown width, cm.

In field measurements, above-ground biomass and below-ground biomass were estimated based on the measured diameter, height, crown diameter, and number of large shrubs (Table 2).

Table 1 Division of baseline carbon strata

Stata no.	Vegetation	Sporadic trees	Natural regeneration	Grazing	Firewood collection	Soil type	Altitude/m	Area/ha
BLS-LX	V-1	No	No	Sporadic	No	Brown soil	1700 2700	626.8
BLS-LX	V-2	No	No	Sporadic	Sporadic	Brown soil	2400 2700	120.0
BLS-MX	V-1	Yes	No	Yes	No	Yellow soil	2400 3000	234.9
BLS-QC	V-1	Yes	Yes	Sporadic	No	Yellow soil	1160 1500	107.1
BLS-QC	V-1	Yes	Yes	No	No	Yellow soil	1100 1700	83.0
BLS-QC	V-2	Yes	No	Sporadic	No	Yellow soil	1330 1800	125.0
BLS-QC	V-2	Yes	No	Yes	No	Yellow soil	1000 1550	164.9
BLS-QC	V-3	Yes	Yes	No	No	Yellow soil	650 1270	120.5
BLS-QC	V-3	Yes	No	No	No	Yellow soil	600 1200	141.7
BLS-QC	V-4	Yes	No	No	No	Yellow soil	1100 1600	136.1
BLS-BC	V-1	Yes	No	No	No	Yellow soil	1400 2400	90.7
BLS-BC	V-2	Yes	No	Sporadic	No	Yellow soil	1400 1900	109.5
BLS-PW	V-1	Yes, mainly Coriaria nepalensis	Yes	No	No	Yellow soil	1300 1540	70.2
BLS-PW	V-2	Yes, mainly Coriaria nepalensis	Yes	No	No	Yellow soil	1360 1880	120.5

V-1: Herbs more than 50% and shrubs less than 10%
V-2: Herbs about 30%, shrubs about 40%, with the height less than 1 m
V-3: Herbs about 20%, shrubs around 50–60%, with the height 1–1.5 m
V-4: Herbal less than 20%, shrubs more than 60%, with the height exceeding 1.5 m

Table 2 Carbon stock of non-tree vegetation before the project

Strata no.	Area/ha	Carbon stock per hectare/(tons of coal · ha^{-1})			Carbon stock/tons of coal		
		Above-ground biomass	Below-ground biomass	Total	Above-ground biomass	Below-ground biomass	Total
BLS-BC	90.7	1.6(0.3)	1.7(0.3)	3.3	145.2	151.0	296.1
BLS-BC	109.5	4.0(1.0)	2.3(0.5)	6.3	438.3	251.5	689.9
BLS-LX	626.8	1.7(0.5)	1.2(0.5)	2.9	1058.2	744.1	1802.3
BLS-LX	121.0	3.5(1.1)	2.2(1.3)	5.7	426.6	265.3	691.9
BLS-MX	234.9	1.8(0.8)	1.4(0.6)	3.3	432.1	337.6	769.8
BLS-PW	70.1	2.8(1.1)	4.7(1.3)	7.5	196.6	330.3	527.0
BLS-PW	120.5	3.9(1.2)	7.0(1.4)	10.8	465.4	838.3	1303.7
BLS-QC	107.1	1.7(0.8)	2.2(0.9)	3.9	182.4	232.7	415.1
BLS-QC	83.0	1.4(0.8)	2.5(0.9)	3.8	115.0	203.6	318.5
BLS-QC	125.0	4.6(1.3)	4.2(1.3)	8.8	572.0	526.0	1097.9
BLS-QC	164.9	4.9(0.9)	3.8(1.2)	8.7	812.8	621.7	1434.5
BLS-QC	120.5	4.9(1.1)	5.2(1.1)	10.0	587.4	620.3	1207.6
BLS-QC	141.7	5.8(1.1)	5.0(1.1)	10.9	825.4	712.6	1538.0
BLS-QC	136.1	8.3(2.6)	5.0(1.7)	13.3	1126.8	683.0	1809.9
Total	2251.8	3.3	2.9	6.2	7384.2	6518.0	13902.2

Data in brackets is standard deviation

The DBH, tree height, age, and crown width of scattered wood were measured within a sample circle with a radius of 20 m to estimate project volume, as well as the above-ground biomass and below-ground biomass (Table 3).

The baseline scenario for justification is to maintain the current status of historical land, namely a wasteland with illegally grazing. Based on the methodology used in this project, the baseline carbon reserve is estimated to be:

For strata currently without any scattered trees, the carbon stock changes for all carbon pools are zero;

For stratum currently without any scattered trees, the change of carbon pools of soil organic, dead and dying trees, and litter is 0. The estimate of the change in carbon stock is based on the number of scattered trees, the growth rate, the allometric growth equation, the IPCC Best Practice Guide, and the default parameters set by regions, countries or IPCC.

(5) Investigate natural regeneration

Because of the distance to the nearby woodland, there are few source seeds that can be spread to the project site.

In spite of the sporadic trees distribution, the current herb and shrub coverage prevent the seeds from falling onto the mineral soil and compete with the seedlings.

Table 3 Carbon stock of trees before the project

Strata no.	Area (ha)	Carbon stock/tons of coal		Total/tons of coal
		Above-ground biomass	Below-ground biomass	
BLS-BC	90.7	8.7	2.0	10.7
BLS-BC	109.5	28.4	6.6	35.0
BLS-LX	626.8			
BLS-LX	121.0			
BLS-MX	234.9	8.3	1.9	10.2
BLS-PW	70.1	0.9	0.2	1.1
BLS-PW	120.5	2.8	0.6	3.5
BLS-QC	107.1	11.7	2.7	14.4
BLS-QC	83.0			
BLS-QC	125.0	18.5	4.1	22.6
BLS-QC	164.9	3.2	0.6	3.9
BLS-QC	120.5			
BLS-QC	141.7	12.9	3.1	16.0
BLS-QC	136.1	7.8	1.8	9.6
Total	2251.8	103.2	23.7	126.9

At least since the 1980s, the land has been without forests, which also shows that natural regeneration is difficult to achieve.

3.8 *Additionality*

Use additional tools to verify whether project activities have additionality.

The project land is forestless before afforestation. The sale of GHG emission reductions is one of the incentive mechanisms for the project to be implemented. If there is no income from the sales of GHG emission reduction, the project is not economically attractive and will face significant obstacles in funding.

Since the government stipulated that the project site should be forest land, while other land uses is not allowed. Current laws and regulations permit the maintaining the existing land use or conducting afforestation/reforestation activities on degraded land, and the most likely baseline scenario is maintaining the conditions of currently degraded land.

1. **Investment analysis**

Select the benchmarking analysis (selection) and use the financial analysis tool from the project concept book developed by the World Bank's BIOCF to analyze the financial internal rate of return (FIRR) without/with carbon benefits, which is 5.36%

Table 4 Figure of sensitivity analysis

	Parameters	Change	FIRR (%)
Without carbon benefit	Product price	+10	6.36
		−10	4.19
	Productivity	+10	6.36
		−10	4.19
	Operating cost	+10	4.81
		−10	5.87
With carbon benefit	Product price	+10	10.22
		−10	8.52
	Productivity	+10	10.22
		−10	8.52
	Operating cost	+10	8.98
		−10	9.84

and 9.42%, respectively (US$5 per tons of carbon dioxide equivalent). The results show that only when there is carbon income can the FIRR exceed the RRR (8%) on forestry investment projects required by NDRC, and the project will be economically attractive.

The sensitivity analysis shows (Table 4) that after a 10% change in the main parameters, the project still has no economic attractiveness in the absence of carbon benefits, and it becomes economically attractive with carbon benefits.[3]

2. **Obstacle analysis**

(1) Investment Obstruction

For farmers, there is no financial mechanism to enable them to invest long term in afforestation/reforestation projects. In the project area, agriculture is the main source of income, but the agricultural product is affected by floods, droughts, and other disasters. Agricultural productivity is very low. The annual average gross income per capita in the project area is US$210, and the lowest in the Bieli Village of Mao County is only US$109. Under this situation, many farmers live below the national poverty line. For farmers, since the revenue from timber and non-wood forest products can only be obtained after the project has been in place for a period of time, they are unlikely to bear the cost of early forestry investment for the afforestation projects. The implementation of the project can earn revenue by selling carbon credits at a relatively early time.

For this type of project activity, no loan can be obtained. It is very difficult to obtain loans from local commercial banks for afforestation/reforestation projects because of its high risk and remote geographical location, which is not economically attractive

---

[3]http://cdm.unfccc.int/methodologies/ARmethodogies/tools/ar-am-tool-02-vl.pdf, www.biocarfund.org.

(for agricultural activities, afforestation where the distance is near and under better conditions can easily to apply for loans for afforestation/reforestation projects). If this project is to be started, there is a higher chance of obtaining a commercial loan.

Through interviews with local communities, it had been found that it is difficult for local farmers and communities to obtain high-quality seedlings, and the quality of their own seedlings is not good enough; it is difficult to support high-quality afforestation activities and to resist fires, pests, and other attacks.

(2) Technical obstacles

The local tree farms have a plenty experience in afforestation/reforestation, but they are usually planted on relatively recent lands with good conditions (easy for planting and forest management) and have no experience in afforestation on remote land plots.

3. **Common practice analysis**

Most of the land in the project is a common property of village and community, but if it is not operated as one economic project, the value of the degraded land and its management will be very low. Therefore, the project site has been left behind and remained degraded after deforestation.

There had been no reforestation activities on similar land in the community. The project of returning farmland to forests that was implemented in China is limited to farmers' slope farmland, government subsidized grain and seeding capital.

The project helps project owners through the assistance of materials and funds to overcome obstacles and financial restrictions.

3.9 Sustainability

Issuing tCER is used for the net anthropogenic GHG removals by sinks generated by this project.

3.10 Estimate of GHG Removals by Sinks

1. **Estimate of baseline net GHG removals by sinks**

Without this project, the existing trees before the project will continue to grow. Stock of changes in above-ground and below-ground biomass carbon stock can be used to estimate the baseline net GHG removals by sinks (Table 5).

2. **Estimate of actual net GHG removals by sinks**

The growth curve was applied to estimate the average stock of each tree species, assuming that the growth curve can be applied to all strata. The growth curve was

Table 5 Estimate of baseline carbon pool change

Year no.	Year	Change of carbon stock for above-ground biomass per year/(tons of coal · year^{-1})	Change of carbon reserve for below-ground biomass per year/(tons of coal · year^{-1})	Annual baseline net GHG removals by sinks/(tons of coal · year^{-1})	Accumulated baseline net GHG removals by sinks/(tons of coal · year^{-1})
1	2007	138.8	31.3	170.1	170.1
2	2008	134.2	30.2	164.4	334.5
3	2009	128.9	29.0	157.9	492.5
4	2010	123.2	27.7	151.0	643.4
5	2011	117.4	26.4	143.8	787.3
6	2012	111.5	25.1	136.5	923.8
7	2013	105.5	23.7	129.3	1053.0
8	2014	99.7	22.4	122.1	1175.1
9	2015	94.0	21.1	115.1	1290.2
10	2016	88.4	19.9	108.3	1398.5
11	2017	83.1	18.6	101.8	1500.2
12	2018	78.0	17.5	95.5	1595.7
13	2019	73.1	16.4	89.5	1685.2
14	2020	68.4	15.3	83.8	1769.0
15	2021	64.0	14.4	78.4	1847.4
16	2022	59.9	13.4	73.3	1920.7
17	2023	55.9	12.5	68.5	1989.1
18	2024	52.2	11.7	63.9	2053.0
19	2025	48.7	10.9	59.6	2112.7
20	2026	45.4	10.2	55.6	2168.3

obtained based on the forest resource inventory data of Sichuan Province once every 5 years since 1990, and some was obtained from published (Table 6).

Use wood density, biomass expansion factor (BEF) and root to stalk ratio (Table 7) to transfer volume to carbon stock. Table 8 shows the carbon stock change of planted trees in each project stratum.[4] Research Institute of Forest Ecology, Environment, and Protection, Chinese Academy of Forestry. China's GHG Emission Inventory—Forestry Section.

3. **Leakage estimation**

Based on the applied baseline and monitoring methodology, there are three potential sources of leakage:

[4]Tianbang Zhao, et al. Study on the growth of magnolia. Journal of Central South Forestry University, 1993 (1).

Table 6 Growth curve of selected tree species of the project

Tree species	Growth curve
Platycladus orientalis	$V = NT(A) \cdot 0.334267111595 \cdot (1-e^{-0.0473470306465 \cdot A})^{3.289912305617}$
Cunning Lanceolata	$V = NT(A) \cdot 0.31225810398 \cdot (1-e^{-0.07402135899476 \cdot A})^{2.527119706049}$
Pinusmassoniana	$V = NT(A) \cdot 0.2903179349968 \cdot (1-e^{-0.07526212201956 \cdot A})^{2.380556958577}$
Quercus acutissima Carruth	$V = NT(A) \cdot 0.1363196248518 \cdot (1-e^{-0.06436380714486 \cdot A})^{2.425820624933}$
Larch	$V = NT(A) \cdot 0.3032245111661 \cdot (1-e^{-0.09092308039843 \cdot A})^{3.84121196496}$
Chinese spruce	$V = NT(A) \cdot 0.33050596446 \cdot (1-e^{-0.029683022346463 \cdot A})^{3.094640905157}$
Pinus tabulaeformis	$V = NT(A) \cdot 0.200346 \cdot (1-e^{-0.054764 \cdot A})^{2.978562}$
Magnolia	$V = NT(A) \cdot 0.151/(1 + 265.2323 \cdot e^{-0.4144 \cdot A})$
Sichuan poplar	$V = NT(A) \cdot 0.3057 \cdot (1-e^{-0.1444 \cdot A})^{2.3398}$
Betula platyphylla and red birch	$V = NT(A) \cdot 0.30583851880298 \cdot (1-e^{-0.08987338464923 \cdot A})^{1.5875471045008}$

Note V, the time is the stock of time t, m^3/ha; NT (A), number of trees, tree/hectare; A, tree age, year

Table 7 WD, BEF, and root to stalk ratio of selected tree species of the project

Tree species	WD/(tons of dry mass · cubic m^{-1})	BEF	Root to stalk ratio
Cunning Lanceolata	0.307 (54,0.009)	1.74 (251,0.08)	0.219 (261,0.004)
Betula platyphylla	0.541 (62,0.018)	1.37 (15,0.04)	0.231 (21,0.018)
Magnolia	0.598 (482,0.012)	1.67 (133,0.04)	0.260 (146,0.010)
Platycladus orientalis	0.478 (32,0.015)	1.77 (45,0.06)	0.244 (50,0.009)
Sichuan poplar	0.378 (144,0.009)	1.59 (108,0.03)	0.291 (107,0.012)
Pinus tabulaeformis	0.360 (15,0.012)	1.62 (74,0.04)	0.231 (89,0.006)
Larch	0.490 (13,0.039)	1.37 (106,0.02)	0.236 (129,0.006)
Pinusmassoniana	0.380 (43,0.019)	1.54 (221,0.03)	0.200 (202,0.007)
Chinese spruce	0.342 (15,0.012)	2.05 (49,0.14)	0.201 (49,0.005)
Quercus acutissima Carruth	0.676 (82,0.012)	1.57 (74,0.05)	0.277 (77,0.009)

Note The data in parentheses represents the number of samples and standard deviation. The parameters of hardwood forest were selected as the parameters of Magnolia

Table 8 Estimate of carbon stock change of planted trees (tons of CO_2/year)

Year	PS-BC-I	PS-BC-II	PS-BC-III	PS-BC-I	PS-BC-II	PS-BC-III	PS-BC-I	PS-BC-II	PS-BC-I	PS-BC-II	PS-BC-I	PS-BC-II	Total
2007	0	0	0	0	1425	0	0	0	71	0	0	1496	
2008	0	152	50	0	1329	0	0	0	36	52	475	4061	6155
2009	777	78	419	189	1903	11	334	9	54	274	1868	7957	14,174
2010	1401	117	748	934	2526	92	1103	57	82	656	3431	10,384	20,529
2011	1692	175	916	1217	3175	260	1729	141	123	1163	4941	12,304	27,627
2012	1841	263	1019	1440	3933	505	2151	250	182	1747	6313	13,304	32,847
2013	1904	391	1093	1625	4484	810	2391	376	269	2361	7513	14,125	37,341
2014	1911	557	1158	1783	5116	1159	2489	511	393	2964	8530	14,637	40,227
2015	1882	842	1229	1918	5719	1538	2481	650	563	3524	9366	14,900	44,613
2016	1828	1207	1321	2033	6286	1933	2401	790	787	4021	10,032	14,961	47,599
2017	1757	1687	1442	2129	6810	2334	2273	926	1062	4441	10,541	14,861	50,263
2018	1656	2277	1593	2207	−7679	2731	2116	1058	1368	−2144	−50,059	14,632	−30,225
2019	1588	2932	1763	221	5405	−1231	−4476	−602	1660	4139	3257	14,301	28,956
2020	1498	3556	1921	1898	5673	2438	1239	910	1872	4446	4532	13,892	43,874
2021	1407	4012	2023	1905	5907	2679	1118	985	1948	4686	5761	13,424	45,856
2022	1317	4173	2030	1903	6107	2903	1002	1055	1862	4858	1035	−9407	18,838
2023	1230	3990	1926	1893	6374	3107	894	1117	−2549	−6997	7353	−26,133	−7895
2024	−6307	−5461	−4981	−2702	6410	3291	793	1174	944	3749	8186	8272	13,367
2025	745	2022	1038	1657	6516	3455	701	1224	730	3770	8871	7875	18,603
2026	691	1565	861	1643	6593	3597	618	1267	540	3756	9413	7477	38,021
Total	18,837	24,554	17,568	24,797	83,813	31,612	21,357	11,897	11,998	41,446	61,355	165,656	514,265

GHG emissions from the burning of fossil fuels in the transportation of seedlings, labor, and logging.

Since the project uses local labor, there is no emission from labor transport. All the seedlings used are from local nurseries or transported from nearby nurseries. Timber felling and timber thinning are required, and the specific transportation method (such as mid-range trucks, agricultural vehicles, livestock, human resources, etc.) depends on the location of the project site. The selected emission factors were 2.6353 kg CO_2/l diesel and 2.49 kg CO_2/l petrol (from the initial national notification). Estimates indicate that during the project crediting period, the leakage was 55 tons of CO_2 equivalent (Table 9).

After project implementation, the carbon stock generated by transferring original grazing and firewood collection to the outside of the project boundary is reducing.

There are firewood collection activities aboveground in this project, and there is illegal grazing on some plots. Therefore, after the project is implemented, these grazing and firewood collection activities will be temporarily or permanently transferred to other areas. When the transferred areas are converted from forest lands to grazing lands, or fuelwood collection activities lead to forest degradation, deforestation,

Table 9 Estimate of project leakage

Year no.	Year	Annual leakage/(tons of CO_2 equivalent · $year^{-1}$)	Accumulated leakage/(tons of CO_2 equivalent · $year^{-1}$)
1	2007	6	6
2	2008	2	8
3	2009		8
4	2010		8
5	2011		8
6	2012		8
7	2013		8
8	2014		8
9	2015		8
10	2016		8
11	2017		8
12	2018		8
13	2019		8
14	2020		8
15	2021	35	43
16	2022	12	55
17	2023		55
18	2024		55
19	2025		55
20	2026		55

vegetation degradation, and other consequences, the transfer of these activities will cause leakage.

Through the participatory rural appraisal, the total amount of livestock processed by project participants inside and outside project boundary and the duration of different livestock grazing was investigated. The analysis shows that the existing grazing land owned by project participants is quite sufficient, and the current grazing intensity is far lower than the bearing capacity. In other words, even if all the grazing activities in the project area are transferred outside the project area, the grazing intensity is still lower than the bearing capacity by then. Therefore, the leakage caused by grazing transfer is zero.

Since any tree felling (firewood collection) at the project site must obtain a government license in advance, the firewood used by the local community for cooking and heating is usually shrubs, dead wood and tree branches. Only a small part of the firewood was collected from the project site. In addition, local farmers can still collect firewood such as shrubs, dead wood, and twigs within the project area, as long as they do not affect forest growth of the project. Therefore, the leakage caused by firewood collection transfer is zero.

The use of wooden fences leads to reduced carbon stock.

The project does not use wooden fences. Therefore, the leakage caused by the use of wood is zero.

4. **Increased GHG emissions**

After project implementation, there will be no significant increase in the emission of other greenhouse gas sources within the project boundary, because:

The soil preparation, thinning and harvest cutting in this project are all done manually and no machinery is used. Therefore, there will be no emissions from fossil fuels burning;

In the process of soil preparation, the method of controlled mountain fire and soil preparation will not be adopted. Therefore, the greenhouse gas emissions from combustion will be zero.

The project does not use nitrogen fertilizer. Therefore, N_2O emissions from the use of nitrogen fertilizers won't be counted.

There is no flood irrigation in the proposed CDM A/R project.

In summary, $GHG_E = 0$.

5. **Estimates of net anthropogenic GHG removals by sinks (Table 10)**

3.11 Main Tree Species and Afforestation Operation

(1) Selected tree species

The tree species selected for the project is shown in Table 11. All species are local, and no genetically modified types.

Table 10 Estimates of net anthropogenic GHG removals by sinks

Year	Baseline net GHG removals by sinks/tons of CO_2 equivalent	Actual net GHG removal by sinks/tons of CO_2 equivalent	Leakage/tons of CO_2 equivalent	Net anthropogenic GHG removals by sinks/tons of CO_2 equivalent
2007	−170	1496	6	- 50,120
2008	−164	6155	2	5989
2009	−158	14,174	0	14,016
2010	−151	21,529	0	21,378
2011	−144	27,627	0	27,483
2012	−137	32,847	0	32,710
2013	−129	37,341	0	37,212
2014	−122	41,227	0	41,105
2015	−115	44,613	0	44,498
2016	−108	47,599	0	47,490
2017	−102	50,263	0	50,161
2018	−95	−30,225	0	−30,320
2019	−89	28,956	0	28,867
2020	−84	43,874	0	43,790
2021	−78	45,856	35	45,742
2022	−73	18,838	12	18,752
2023	−68	−7895	0	−7963
2024	−64	13,367	0	13,303
2025	−60	38,603	0	38,543
2026	−56	38,021	0	37,965
Total	−2168	514,265	55	460,603

(2) Technology adopted by the project

The Project implementation strictly complies with the following technical standards:

- National Afforestation Technical Regulations (GB/T15776);
- National Standards for Non-commercial Forest Construction (GB/T18337.1, GB/T18337.2, GB/T18337.3);
- Design Code for Afforestation Operation (LY/T1607-2003);
- Regulation of Techniques for Comprehensive Control of Soil Erosion (GB/T16453.1-16453.6);
- Regulations for Forest Tending (GB/T15781);
- Classification of Forest Tree Seed Quality (GB7908);
- Tree Seedling Quality Grading of Major Species for Afforestation (GB6000);
- Technical Regulations for Cultivation of Tree Seedling (GB/T6001);
- Technical Regulations of Containerized Seedling (LY1000).

Table 11 Tree species selected for the project

Tree species	Area (ha)					Total (ha)
	Beichuan county	Li county	Mao county	Pingwu county	Qingchuan county	
Platycladus orientalis		109.0				109.0
Betula platyphylla	56.4				274.1	330.5
Sichuan Poplar			63.0			63.0
Red Birch	28.0	34.4				62.4
Magnolia	115.8			40.6		156.4
Larch				120.0		120.0
Quercus acutissima Carruth				30.0	264.2	294.2
Pinusmassoniana					66.0	66.0
Minjiang cypress (2007)		358.0				358.0
Minjiang cypress (2008)		109.0				109.0
Cunning Lanceolata					274.0	274.0
Pinus tabulaeformis		137.4	85.9			223.3
Chinese spruce			86.0			86.0
Total	200.2	747.8	234.9	190.6	878.3	2251.8

(3) Species arrangement

Afforestation is performed using a block-mixed model (where the same species are planted in each subplot) to reduce risk (fire, pest, and disease) and increase environmental and social benefits.

(4) Seedling and nursery

The seedlings used in the project are local seedlings. The seeds needed for seedling raising are collected from local mother trees with similar site conditions or local seedling bases. All seeds have to be certified, pass quality inspection, and come with qualification label. Each batch of the seeds must have a quality certificate to indicate its seed provenance and quality grade. The quality certification of seedlings is carried out according to the national standard (GB6000-1999). All seedlings used have to reach the quality standards at a minimum.

Platycladus orientalis, *Minjiang cypress*, *Pinus tabulaeformis*, *Pinusmassoniana*, and *larch* will be grown in nutritious bags for their seedlings. The method is to grow seedlings in plastic bags (5 cm in diameter and 15 cm in length) containing 20% organic soil. This technique guarantees the growth conditions at the initial stage of planting and consequently improves the survival rate and facilitates early growth.

Some *Minjiang cypress* will use large seedlings for transplanting; *Quercus acutissima Carruth* will use the point-planting method, that is, the seed will be directly planted; *Sichuan Poplar* will use the cutting method for the seedling.

(5) Soil preparation

In order to prevent soil erosion and protect existing carbon pool, the project prohibits land burning and whole reclamation site preparation, instead, it adopted stripe-wide removal (with a width of 100 cm) of the existing non-tree vegetation along the contour line. In autumn and winter, burrow-shaped soil preparation is performed on the cleared belts (*magnolia, larch, pinusmassoniana, platycladus orientalis,* and *cunning lanceolata* follow a diameter of 40 cm and depth of 30 cm; *red birch, betula platyphylla*, and *Minjiang cypress* follow 60 cm in diameter and 50 cm in-depth; *Quercus acutissima Carruth* uses plowed burrow of 30 cm in diameter and 20 cm in-depth; cutting *Sichuan poplar* uses pits of 2–3 cm in diameter). All pits and burrows are distributed on the sloping land in a triangle shape.

(6) Afforestation

Afforestation will last for 2 years, beginning from the spring of 2007. Afforestation activities were mostly conducted during rainy days, and *Quercus acutissima Carruth* were planted by hand. Other tree species are planted manually.

The afforestation density of each tree species is as follow:

Magnolia, Sichuan poplar, betula platyphylla, and red birch: 2 m × 3 m, 1,667 trees per hectare;

Quercus acutissima Carruth: 1.5 m × 1.5 m, 4,444 trees per hectare;

Cunning lanceolata, platycladus orientalis, Minjiang cypress, pinus tabulaeformis, Pinusmassoniana, larch, and Chinese spruce: 2 m × 2 m, 2,500 trees per hectare.

At the initial stage, in order to ensure a high survival rate and better growth status, weeding and soil loosening will be performed twice a year for the first three years. No weeding and loosening will be done during the rainy season.

(7) Forest Management

According to the natural thinning rate, different carbon strata and growth of tree species, thinning was carried out in the woods for 15–40 years. The thinning intensity was 20–30% (including the natural thinning rate) of timber stock and tree density. The specific thinning standard is based on specific tree species and growth. The purpose of forest management is to promote forest growth and obtain more wood. From the perspective of ecological environment of the river basin and the prevention of water and soil erosion, the project will not perform clear-cutting of the new afforestation in Mao County. Other land plots will start to harvest cutting after 10–80 years (Table 12).

After the forests have been harvested, they will be renewed through plantation afforestation. Plantation means is adopted and no machinery will be used, whether it is the harvest cutting, thinning or replanting.

Table 12 Forest management

County	Species	Thinning time/year	Thinning intensity (percentage of stock volume)/(percent)	Harvest cutting time/year
Beichuan	*Betula platyphylla*	15,25	30	30
	Red Birch	15,25	30	30
	Magnolia	15,25	30	30
Li	*Red Birch*	40	20	80
	Platycladus orientalis	40	20	80
	Minjiang cypress	40	20	80
	Pinus tabulaeformis	40	20	60
Mao	*Sichuan poplar*			No harvest cutting
	Pinus tabulaeformis			No harvest cutting
	Chinese spruce			No harvest cutting
Pingwu	*Quercus acutissima Carruth*			10
	Magnolia	15,25	30	30
	Larch	15,25	30	30
Qingchuan	*Betula platyphylla*	15,25	30	30
	Quercus acutissima Carruth			10
	Pinusmassoniana	15,25	30	30
	Cunning Lanceolata	15,25	30	30
Beichuan	*Betula platyphylla*	15,25	30	30
	Red Birch	15,25	30	30
	Magnolia	15,25	30	30
Li	*Red Birch*	40	20	80
	Platycladus orientalis	40	20	80
	Minjiang cypress	40	20	80
	Pinus tabulaeformis	40	20	60
Mao	*Sichuan poplar*			No harvest cutting
	Pinus tabulaeformis			No harvest cutting

3.12 Investment Overview

Afforestation activities are planned to start in the fall of 2007 and the summer of 2008. The estimate of activity cost is based on the standards for natural forest conservation program and the Program of conversion farmland to forests, the afforestation standards of Sichuan Province, market information issued by provincial and local governments in 2006, and local socioeconomic and market conditions. The total investment of the project is RMB28.38 million, of which the construction period (the first three years) will take RMB21.73 million and the operation period RMB6.65 million. Project development costs (baseline investigation, PDD preparation, validation, and verification) were not taken into account.

3.13 Key Elements of Monitoring

1. **Monitoring content and method**

(1) Monitoring Project Boundary

After afforestation, a field survey is conducted on the actual project boundary of each plot using remote sensing or GPS to check whether the actual project boundary is consistent with that described in the PDD.

The geographic location that is consistent with the PDD is entered into the GIS system and the area of each layer and stand is calculated.

During the project crediting period, if there is any change in the forest land, such as deforestation within the project area, clear records should be made of the area and location of deforestation. Similarly, if a plot within the project boundary is planted.

(2) Monitoring of afforestation activities

Compared with the PDD, the following monitoring activities will be conducted in the first three years after afforestation:

- Mountain burning: time, place, area, biomass consumption, combustion, and other methods used;
- Site Preparation: time, place, and area of each layer;
- Plantation: time, place, area, tree species (establishment of stand type), and quality of seedlings;
- Fertilization: time, place, area, species, and amount of each fertilizer used (first 3 years);
- Survival rate: The first survival rate inspection should be conducted after 3 months of the plantation. If the survival rate is lower than 85% of afforestation density, replanting activities should be carried out;
- Preservation rate: Final inspection after 3 years of planting;
- Weeding: check whether Is it consistent with the description of the PDD;

- Record any issues that do not match with the operation plan of the PDD.

(3) Forest Management Monitoring

The monitored forest management activities include:

- Thinning: time, place, area, species, thinning intensity, volume, or biomass that were thinned;
- Harvest cutting: time, place, area, tree species, volume, or biomass of harvest cutting;
- Weeding: time, place, area, tree species, volume, or biomass;
- Firewood collection: time, place, area, tree species, volume, or biomass;
- The renewal of cutting site;
- If natural renewal happened on the cutting site, check the conditions of natural regeneration;
- Monitoring of various disturbances: time, place, area (GPS coordinates or remote sensing), tree species, type of disturbance, loss of biomass, countermeasures, and change of project layer and stand boundary.

(4) Monitoring of net GHG removals by sinks

The selected methodology AR-AM0003/03 does not require the monitoring of net GHG removals by sinks. However, extending the crediting period requires the collection of new data to establish a new baseline. After collecting GHG sinks removal during the crediting period, it will be determined whether the baseline method and the baseline scenario are still valid and whether they need to be updated.

At the end of the project crediting period, above-ground biomass can be measured to estimate carbon stock changes in the baseline scenario. At least three sample plots are randomly selected from outside the project boundary for measurement, which is used to represent the degraded land when there is no afforestation/reforestation project. The accuracy level is 10% and confidence level 95%.

(5) Monitoring of actual net GHG removals by sinks

In order to monitor the change of carbon stock for both above-ground and below-ground biomass within the project boundary, it is needed to measure factors of each sample plot based on the result of carbon layer stratification. The data needed for monitoring carbon pool change within the project boundary mainly include:

- area of each carbon stratum;
- number of tree species in each sample plot;
- age of each tree species in sample plots;
- the DBH and tree height of standing and deadwood in the sample plot;
- use locally available formula to calculate the volume:
  - Pinus tabulae form is: $V = 0.000066492455 \cdot DBH^{1.8655617} \cdot H^{0.93768879}$
  - *Pinusmassoniana*: $V = 0.000060049144 \cdot DBH^{1.8719753} \cdot H^{0.97180232}$
  - Minjiang cypress and *Platycladus orientalis*: $V = 0.000057173591 \cdot DBH^{1.8813305} \cdot H^{0.99568845}$

- *Quercus acutissima Carruth*: $V = 0.000059599784 \cdot DBH^{1.8564005} \cdot H^{0.9956206}$
- *Betula platyphylla*: $V = 0.000048941911 \cdot DBH^{2.0172708} \cdot H^{0.88580889}$
- *Cunning Lanceolata*: $V = 0.000058777042 \cdot DBH^{1.9699831} \cdot H^{0.89646157}$
- *Chinese spruce*: $V = 0.000056790543 \cdot DBH^{1.8517320} \cdot H^{0.334624}$
- *Larch*: $V = 0.000068320 \cdot DBH^{1.74136} \cdot H^{1.11525}$
- Other hardwood species: $V = 0.000052750716 \cdot DBH^{1.9450324} \cdot H^{0.93885330}$

Among them, V, volume, cubic meters/strain; DBH, DBH, cm; H, tree height, m.[5]

The total volume of all trees in the sample plot is added up to obtain the total stock (m^3/ha) per unit area. The total stock per unit area of tree species *j* in the *i* layer can then be calculated.

(6) Leakage

According to the applied methodology, there are three sources of leakage that may be caused by the project: fossil fuel consumption caused by vehicle, displacement of grazing and firewood collection activities, and use of wooden fences. Since the leakage of grazing and firewood collection activities and the use of wooden fences are both zero, the only source of leakage that needs to be monitored is the fossil fuel consumption caused by vehicles, mainly the transport of labor and the use of vehicles in the production process. When monitoring, the vehicle type, fuel type, mileage, and fuel consumption should be recorded.

The estimation of GHG emissions from vehicle use will use national default parameters in the national GHG inventory. For example, the emission factor for diesel or gasoline used by local farmers is 2.6353 kg CO_2/L diesel or gasoline. The diesel consumption is generally 0.1–0.2 L/km, and the fuel consumption of different vehicles varies.

2. **Sample setting**

(1) Post-stratification

Because project implementation results may be inconsistent with the PDD, especially planting species, time, boundaries, etc., plus that there are various unforeseen disturbances (for example, fire and pest outbreaks, etc.), each monitoring needs to be carefully evaluated to identify whether there is a need to reclassify pre-demarcated carbon strata.

(2) Distribution of sample and plots

The established permanent sample plots should be measured regularly. Assuming that the cost of establishing sample plots within each stratum is constant, using the formulas (62) and (63) in Sect. 2.2.1 of the methodology to calculate the number of sample plots within each layer (Table 13). The volume of mature trees (Q, cubic meters per hectare) was calculated based on the growth formulas. The standard deviation (*sti*) of each layer is set at 20%, and the accuracy level of 10%. In order to

Table 13 Number of monitored sample plots per stratum

Layer no.	Species	Time	Area/ha	Number of sample plots
PS-BC-	*Betula platyphylla + Red birch*	2008 Spring	56	3
PS-BC-	*Magnolia*	2008 Spring	87.4	3
PS-BC-	*Magnolia + Betula platyphylla*	2008 Spring	56.8	4
PS-LX-	*Pinus tabulaeformis + Red birch*	2007 Spring	171.8	4
PS-LX-	*Minjiang cypress*	2007 Spring	358	8
PS-LX-	*Minjiang cypress + Platycladus orientalis*	2008 Spring	218	5
PS-MX-	*Sichuan poplar*	2008 Spring	63	3
PS-MX-	*Chinese spruce + Pinus tabulaeformis*	2008 Spring	171.9	3
PS-PW-	*Magnolia*	2007 Autumn	40.6	3
PS-PW-	*Larch + Quercus acutissima Carruth*	2007 Autumn	150	5
PS-QC-	*Pinusmassoniana + Quercus acutissima Carruth*	2007 Autumn	330.2	7
PS-QC-	*Cunning Lanceolata + Betula platyphylla*	2007 Autumn	548.1	21
Total			2251.8	69

ensure statistical independence of all layers, the minimum number of sample plots per layer is set at 3.

After the first monitoring, the amount of sample plots can also be adjusted based on changes in carbon stock and disturbances.

(3) Sample size

The density of stands is approximately 1,600–2,500 plants/ha, so the plot size is set at 400 m^2 (approximately 64–100 per plot).

(4) Plot location

The location of the permanent sample plot will be randomly set.

[5]Forestry Bureau. 1984. Sichuan forestry inventory—harvesting tables and growth formula.

(5) Measurement and estimation of changes in carbon stock

Each monitoring will measure the growth of each tree in the plot. For conservative considerations, the trees that existed before the start of the project are not measured and calculated. The methodology used also does not require the measurement and calculation of non-tree vegetation (i.e., herbs, grass, and shrubs). Biomass expansion factor (BEF) was used to estimate the change of carbon reserves in live standing trees in each plot.

(6) Monitoring of GHG emission sources caused by the project

As there is no controlled mountain fire and full-scale soil preparation, no nitrogen fertilizer used, no flood irrigation adopted and no machinery during the operation, this project will not bring significant sources of GHG emissions.

3. **Monitoring frequency**

The project afforestation began in 2007. *Betula platyphylla*, *red birch*, *magnolia*, *larch*, *pinusmassoniana* and *cunning lanceolata* will be thinned at the 15–25 years, and Quercus acutissima Carruth will be cut after 10 years. In order to avoid coinciding with the peak of carbon stock, the first monitoring will be conducted in 2012 and will be monitored once every five years (2017, 2022, 2027, etc.). Project verification will be carried out in 2012, 2017, 2022, and 2027, respectively (Table 14).

4 Project Implementation

4.1 Main Activities

1. **Project site selection investigation**

In order to ensure that the project achieves the expected objectives and serves as a pilot, the Forestry Department of Sichuan Province organized the officials of relevant departments and experts to discuss and determine eight principles for the selection of FCCB projects, namely:

- Has certain representative features (including landform type, vegetation, plantation type, social and economic conditions, etc.);
- Hotspots or neighboring sites of the nature reserves that are of concern to Conservation International (CI) or other international agencies;
- Ecologically fragile zones that is rich in biodiversity, or areas that are important for the conservation of biodiversity;
- Area where the plantation size is relatively large and has a certain scale of suitable forest lands;
- Ability to undertake projects (including enthusiasm, technical strength, experience in international cooperation projects, monitoring conditions);

Table 14 Time of monitoring, harvest cutting, and thinning

Year no.	Year	Monitoring	Verification	Cutting	Thinning
1	2007				
2	2008				
3	2009				
4	2010				
5	2011				
6	2012	X	X		
7	2013				
8	2014				
9	2015				
10	2016			X	
11	2017	X	X		
12	2018				
13	2019				
14	2020				
15	2021				X
16	2022	X	X		
17	2023				
18	2024				
19	2025				
20	2026				X
21	2027	X	X		

- Can be combined with other projects (for example, artificial pilots of commercial logging);
- Areas where the basic data are relatively complete and have research basis;
- Convenient transportation, with Chengdu as the center and within a half-day journey.

Through expert discussions, the project selection process is drafted, county-level basic information investigation form is prepared, so do county-level information investigation and evaluation form, village-level information investigation form and basic information investigation form of the nature reserve. Experts are organized to conducted field visits in Li County, Mao County, Wenchuan, Ya'an, Tianquan, Baoxing, Qingchuan, Pingwu, and Chaotian districts and other nature reserves such as Wanglang and Tangjiahe, so as to collect basic data, and provided scientific basis for unifying project goals and strategies, and further improving project selection. At the same time, an FCCB project workshop was held with the participation of leaders and experts from the State Forestry and Grassland Administration, Forestry

Department of Sichuan Province and The Forestry Department of Yunnan Province, with the purpose of discussing project questionnaires, selection process, and work lists. The workshop has positive guiding for the project selection and also provides useful references for similar projects in other regions.

In accordance with the project process and the principle of candidate county selection, the afforestation office and conservation office of the Forestry Department of Sichuan Province and the forestry resources monitoring center of the Sichuan Forestry Inventory and Planning Institute utilized three indicators, namely the distribution of forest resources, the distribution of nature reserves, and the status of barren hills suitable for afforestation and the cultivated land on steep slopes, to jointly analyze the importance of biodiversity and the scale of suitable forest land. A preliminary selection of 43 candidate counties was selected, and a training course on basic information application for forest carbon projects in candidate counties was conducted. Through analyzing the data compiled by candidate counties, 10 were selected as the target of the field visit. The focus of field visit is as follow: contribution to biodiversity, whether it is land suitable for forests and cultivated land on steep slopes that connect with fragmented habitats, corridors, water sources or within major river basins; whether the scale is relatively concentrated and contiguous, and whether it meets the requirements of implementing carbon sink projects (>1000 μm). Through expert investigation and discussions, Qingchuan County, Pingwu County, Beichuan County, Li County, and Mao County were selected as pilot project counties.

2. **Prepare the design guidance on project investigation**

Because the past technical guidance on afforestation planning and design only focused on background investigation and design, without considering the factors of biodiversity and community, it can't meet the FCCB requirements. To this end, experts from the Forestry Department of Sichuan Province and CI began to prepare a background investigation and afforestation design guidance that in consistent with FCCB projects. Through several rounds of discussions and three major revisions, the "Framework Guidance on Project Background Investigation" and "Framework Guidance on Project Design" were drafted. Experts of forest carbons, ecological restoration, biodiversity conservation, and community development, coming from CI, Chengdu Institute of Biology of China Academy of Sciences, Sichuan Academy of Social Sciences, Sichuan Forestry Science and Research Institute, Sichuan Forestry Inventory and Planning Institute and related offices of Forestry Department of Sichuan Province, conducted discussions on the structural integrity, logic, science, practicality and background investigation indicators of the framework. After the first draft was completed, the engineers and technicians of Sichuan Forestry Investigation and Design Team carried out verification based on the design guidance and submitted a user's report. Experts of the draft team made further amendments and improvements to form an *Investigation and Design Guidance on CDM Afforestation and Reforestation Carbon Sink Project (trial version)*, which was issued as a textbook for province-wide training in forest carbon knowledge. It provides guidance on background investigation and project design of forest carbons in Sichuan Province.

3. **Conduct project background investigation**

With Sichuan Forestry Inventory and Planning Institute and Sichuan Forestry Science and Research Institute as technical support, initial information and carbon baseline investigation were conducted on the primary project plots to form up the *Initial Information Investigation Report on CDM A/R Projects in Northwest Sichuan, GIS Data Report of CDM A/R Project in Northwest Sichuan, Carbon Baseline Investigation Report on CDM A/R Project in Northwestern Sichuan* and *Afforestation Operational Design of CDM A/R Project in Northwest Sichuan*. With the technical support of the Institute of Rural Economics of the Sichuan Academy of Social Sciences, a socioeconomic investigation and assessment of communities covered by the project was conducted to prepare the Community Investigation and Assessment Report of CDM A/R project in Northwest Sichuan.

(1) Training and discussion on investigation approach

In order to standardized project development methods and ensure the quality of project development, training seminars on forest carbon project development methods, CCB standards and community assessment methods, and carbon baseline investigation and practice of FCCB projects were held in succession. Nearly 100 people from Forestry Department of Sichuan Province, Sichuan Forestry Inventory and Planning Institute, Sichuan Forestry Science and Research Institute, CI, TNC and project owners and implementing counties participated in related training seminars. The content covers CDM process, the method for drafting afforestation/reforestation project summary (PIN), material collection for CDM A/R PDD, CCB standards, participatory rural appraisal, and carbon baseline investigation procedures and methods.

(2) Examine the eligibility of project site

A total of 10 persons from the forest carbon office of Forestry Department of Sichuan Province, experts invited by CI, and provincial personnel engaged in implementing forest carbon project conducted a field study with regard to the proposed carbon projects in Pingwu, Qingchuan, Beichuan, Mao County, and Li County. Through the study, guiding opinions on project boundary determination and carbon stratification criteria were put forward, and the baseline investigation methodologies to be adopted by these 5 counties were identified. This serves as the guidance for the next-step work and lays the foundation for baseline investigation. Meanwhile, the project counties also get a more intuitive understanding of the project site.

(3) Measure carbon baseline

The purpose of measuring the carbon baseline is to calculate the basic carbon of the project site and predict the GHG leakage caused by project activity. Due to altitude, latitude, slope-side, and degree and climate, there are great differences in plant types, proportion of various plants, plant growth status, and grazing conditions in the project counties, thus carbon stratification should be carried out by counties. For each sample plot, it is needed to investigate all plant biomass, including herbs, shrubs, scattered trees, and natural regeneration. The sample plots were determined

Table 15 Number of sample plots in each baseline layer

Counties	Number of sample plots, subplots				
	Sample plots	Herb subplots	Shrub subplots	Large-scale shrub subplots	Round tree plots
Beichuan	26	52	52		26
Li	65	260	4	1	65
Mao	20	80	10		20
Pingwu	45	104	98		45
Qingchuan	61	157	197	20	61
Total	217	653	361	21	217

by a random method, using Random points selectors in Arcview3.x for random site placement. See Table 15 for the number of sample plots and subplots in each county.

(4) Afforestation and forest management design

First, the forestry investigation and design team of Sichuan Province organized engineers and technicians to form up a project baseline investigation and design team. After the training, they went together with CI staff and personnel from the carbon office of Forestry Department of Sichuan Province into Mao County, Qingchuan and Wanglang for field investigations. They completed the baseline investigation report, operation design and validation of the *Investigation and Design Guidance on CDM Afforestation and Reforestation Carbon Project (trial version)*. After the adoption of the *Design Guidance on CDM Afforestation and Reforestation Carbon Project (Trial version)* and in accordance with the new design guidance, Mao County, Qingchuan and Wanglang finished preliminary investigation and design. Engineering and technical personnel from 5 counties conducted an investigation of baseline information, carbon baseline measurement, community surveys and afforestation operations. A total of 36 project sites in 29 villages of 21 townships in 5 counties of northwestern Sichuan were investigated and 89 subplots were classified, covering an area of 2,251.8 ha. There would be 12 species to be planted, and the afforestation was expected to complete in 2 years. The initial plantation is estimated at 5.39 million.

(5) Participatory rural appraisal

The participatory rural appraisal was conducted in parallel with the baseline investigation and project design, and was completed by a rural appraisal team composed of experts from Sichuan Academy of Social Sciences and engineering technicians from the County Forestry Bureau. In the process of community investigation, a variety of participatory investigation tools were used, such as semi-structured interviews, community meetings, community resource maps, farming season calendars, memorabilia and matrix rankings. The focus is to have discussions with the head of the village, village Party secretaries, village cadres and other key stakeholders, and to conduct random interviews with 10–15 households of different types in each village. A total of 348 farmer households in 29 villages of 5 counties, were visited to understand the current situation of community social economy, land use, forest land ownership, firewood and other energy use and grazing activities, as well as farmers' willingness to participate in the project and their understanding of the project. The villagers' meeting was also held in each village, which introduced the carbon project, understood the main problems and demand of the community. The meeting also intended to facilitate villagers to negotiate on different ideas and try to find solutions that can contribute to consensus. Meanwhile, the meeting helped to obtain suggestions from different interest groups on project implementation and sorting tree species preferences. The investigation shows that local villagers can recognize the environmental benefits brought by the forest and have a strong sense of conserving rare and endangered animals and plants. However, farmers have little awareness of forest carbon and they have never heard of it. After the introduction, villagers expressed a strong desire to participate in the project, believing that through this project, they can find a job not far away from home. They can get income while taking care of farmland and livestock; obtaining more income by selling CERs; Greening barren hills and wasteland can improve the environment, protect farmland, reduce droughts, floods, landslides, and other natural disasters; they can also gain more experience in planting trees and forest management by participating in technical training. However, many farmers feel that carbon trading is invisible and intangible, and they are skeptical of their authenticity. The agreement reached in project implementation is that the farmers provide land and the operating entity invests in afforestation.

(6) Biodiversity investigation

The investigation included collecting, collating and analyzing second-hand information and filed investigation on biodiversity in the project area. The baseline field investigation was conducted in conjunction with a carbon baseline investigation. The investigation included plants, mammals, rodents, amphibians, reptiles, and birds. Investigations on biodiversity and species of endangered plants were conducted on different carbon strata. Traps and fence traps were used to investigate the number and types of amphibians, reptiles, and rodents. Two investigation lines are set randomly within each project block to investigate large mammals and birds, traces of footprints, feces, and individuals were investigated and recorded to identify the types of these mammals and birds. The analysis shows that the project areas are all in the nature reserves and their neighboring communities, with extremely diverse biodiversity. However, field study showed that the project sites had low biodiversity and there were no animals and plants under national protection or IUCN-listed endangered species.

(7) Evaluate environmental impact

Due to severe human disturbances that have persisted in the past, such as continuous deforestation, excessive firewood collection, and illegal grazing, many of the project lands had been seriously degraded and soil erosion was very serious. If the current situation is not changed, these lands will be further degraded. Therefore, the implementation of the project will contribute to soil erosion and reducing natural disasters such as landslides. The project counties completed the environmental impact registration forms and submitted them to local environmental protection bureaus for approval. The registration forms stated that the main purpose of the projects was to restore the degraded land.

4. **Sign a cooperation agreement**

The Dadu River Afforestation Bureau has reached a tripartite cooperation agreement with county forestry bureaus and forest landowners through several consultations. The "Cooperative Afforestation Contract for Forest carbon Project in Northwestern Sichuan" was signed to determine the rights and responsibilities of the three parties.

5. **PDD Preparation**

In May 2007, project counties completed supplement and improvement of related contents such as grazing conditions and harvesting and logging. The Sichuan Forestry Inventory and Planning Institute completed the growth formula of afforestation tree species. In July 2007, the PDD was completed. Afterward, it was revised in accordance with changes in the new template of PDDs, and the fourth edition, which was officially submitted, was publicized on January 20, 2008. Subsequently, based on the review of the NDRC and the DOE's verification opinion, it was revised again and the final 6th edition was released on June 11, 2009.

6. **Project application and approval**

This project was reviewed by the Forestry Department of Sichuan Province and approved by the State Forestry and Grassland Administration. After being reviewed at the 52nd meeting of the National CDM Project Review Board, the project was certified by the NDRC. The Dadu River Afforestation Bureau was authorized as the implementing body for conducting project activities.

7. **DOE validation and CDM-EB registration**

In mid-April 2008, TüV-SüD, as a DOE authorized by the UN's CDM-EB, verified the project's eligibility. Through the review of the PDDs and on-site investigation and verification of certain afforestation plots in 4 counties of Li County, Mao County, Qingchuan County, and Pingwu County, DOE officials affirmed and highly praised the prospects of the project and the initial work of carbon project. Meanwhile, suggestions and opinions were put forward for revising PDDs and project operation. After improving the PDDs, the TüV-SüD reviewed and adopted the 6th edition of the PDDs and officially issued a validation report on June 15, 2009. After the CDM-EB review, online notification and other procedures, the CDM-EB under the UNFCCC approved the project registration on November 16, 2009.

8. **Project operation**

(1) Afforestation

The project's soil preparation and afforestation activities are carried out by county forestry bureaus or management office of Nature reserves. The counties entered the seedling preparation stage in September 2006 and the project was launched in full scale in March 2007. The project implementation mainly depends on the operation teams organized by villagers in the neighboring communities. The nature reserves are responsible for the technical training before the construction. During the implementation, technical personnel and other staff are sent to provide on-site guidance and quality supervision. All counties identify afforestation contractors or farmer households for replanting, tending and management work, thus the overall afforestation outcome is relatively good. However, in some plots, the survival rate of seedlings is below 85% and even lower, and there are still a few plots where all the seedlings are dead, and the severe stampede by cattle and sheep.

(2) Community Projects

In order to demonstrate the experiments and pilots of community participation and benefits in FCCB projects, pilot projects for edible mushroom cultivation and biomass solid fuel processing and utilization were established in Qingchuan and Pingwu, involving 4 villages and 1 nature reserve. The project adopts the principle of self-reliance on a voluntary basis. The funding is mainly from project funds, supplemented by farmers with their own financial resource. A total of more than 30 gas stoves, 7 biogas pools, and 3 energy-saving stoves were built to cultivate 19,000 bags of edible fungi. The project also invited experts from Sichuan Academy of Agricultural Sciences and local farmers to provide technical training and guidance. A "calendar of bagged mushroom cultivation" was prepared and distributed to villagers.

9. **Carbon trading**

After negotiations with the Hong Kong Carbon Care Asia, at the end of November 2007, the Dadu River Afforestation Bureau, the Hong Kong Carbon Care Asia, and the Beijing Shanshui Conservation Center successfully signed the first verification period (2007–2012) of the project. The cooperation agreement covered approximately 50,000 tons of CO_2 emission reduction trade, which is unprecedented for Sichuan's forestry sector, it shows that carbon trade has entered a substantive stage and moved into the international market.

10. **Project thematic study**

(1) Project comprehensive database development

In order to store, manage and use a large amount of basic data collected by FCCB project, the project developed a comprehensive database of forest carbon projects, which kept data collected from more than 30 potential project counties, which can provide support for the development of same type projects.

(2) Study on biomass models of shrubs

Among the project plots in the pilot area, four major shrub species such as coriaria nepalensis, emei rose, wild pepper (temporary name), and Berberis kawakamii Hayata. In order to quickly and accurately measure the biomass of shrub species, and find out the baseline status of main shrubs in the region for future predictions, the project contracted with Sichuan Academy of Forestry to complete a study on the estimation model of the biomass of the four shrubs.

4.2 Difficulties and Solutions

The risks and difficulties encountered during project implementation and countermeasures are as follow:

(1) Project fundraising

This project is a unilateral project of CDM A/R. 3M, CI, TNC, and the Beijing Shanshui Conservation Center supported the project development. However, the project preparation and operation need to be financed by the owner. In particular, there was a lack of workforce after the earthquake, thus labor costs increased and project investment cost was far greater than expected. To this end, the project owners, county forestry bureaus and nature reserves actively reported to Forestry Department of Sichuan Province and the local governments, receiving strong support to ensure funding for the operation period. After the fund is in place, it is necessary to further coordinate the funds for future operation so that the project management and monitoring can be smoothly conducted.

(2) The impact of natural disasters

Beichuan County, Qingchuan County, Pingwu County and Mao County were the hardest-hit areas of the "5·12" Wenchuan Earthquake. Li County was the hardest-hit area. The afforestation plots, forest roads and project communities all suffered severe destruction, and the project is very hard to move forward after the natural disaster. Due to geological disasters such as landslides, the damaged surface of some project sites requires a large amount of investment to recover, coupled with the shortage of labor after the earthquake, the afforestation and replanting work was postponed and generally completed until 2010.

(3) Forest land and wood ownership

The collective forest ownership reform results in the inconsistency between the operating rights of individual plots and the contracts signed at the time of project development. This needs to be re-coordinated and clarified.

(4) Management of planted forest land

The implementation of the project may also have negative effects in certain aspects, which will have a negative impact on the utilization of resources and the livelihood of local farmers. For example, the implementation of the project occupies the original barren hills and wasteland, which affects firewood collection, so that villagers may have to go farther to collect firewood; at the same time, the project may have a significant impact on grazing. In order to maintain the original number of livestock, it may be needed to keep them in fences or directly reduce their number. These will have to be resolved through direct consultation with communities and through community engagement in drafting a practicable covenant.

(5) The potential threat of fire

Although the frequency of fire in the project communities is little, most of the project sites are adjacent to community farmland, which may lead to increased fire threat after afforestation. The project intends to reduce the risk of fire through technology and awareness training of for local farmer households and communities, strengthening of management and monitoring, establishment of fire barriers and creation of mixed forests.

5 Project Benefits

5.1 Society

(1) Increase awareness of the FCCB and forest carbon

In the past, government officials and villagers are mainly interested in the benefits of forests wood products and the role of forests in maintaining water and soil and in environmental improvement. The project brought new ideas such as forests with multiple benefits and carbon, enriched the content of traditional ecological benefits, and made people understand what kinds of projects have multiple benefits through the promotion of CCB standards. The background investigations, baseline measurement and community assessments of carbon projects have diverse content and stricter requirements than traditional forestry planning and design. This has also brought new inspiration to traditional forestry and posed new challenges.

(2) Incubate a talent team for the development and implementation of forest carbon projects

The capacity building of forest carbon project is the key elements of the project for three years. Ten seminars and training had been conducted with about 500 participants. The participants include officials and technicians from forestry authorities at all levels and provincial-level research institutes, colleges and universities, and enterprises and public institutions. The contents range from basic knowledge of forest carbon to project development technologies. Some engineering technicians, through seminars, training, and practical activities, mastered the technical points of project development and became the backbone to sustain the project, which provided conditions for developing forest carbon project in the province. Liangshan Prefecture and Panzhihua City also completed the development of the project concept document, which laid foundations for future projects.

(3) Enhance community cohesion

Through participatory methods, the project allows cadres and technicians to move into the community and get closer to people to fully listen to the opinions of community villagers. Participatory tools are used to allow community villagers to participate extensively in the planning and decision-making of the project, thereby enhancing villagers' self-confidence and participation awareness. Meanwhile, the project also organized farmers of decentralized production to communicate and learn from each other through community meetings, technical training, and centralized construction, thus enhancing the affinity of government and the cohesion of the community.

(4) Improve community villagers' skills

Community investigation shows that villagers lack certain technologies in forest development, management, and protection. The project provides systematic training for villagers through local forestry authorities in accordance with seasons and procedures, so that they can master the techniques and skills of seed selection, soil preparation, planting, tending, and comprehensive management of forest pests and diseases.

5.2 Economy

The project community investigation shows that the per capita annual gross income in the region is US$210. The lowest one—the Bieli Village of Mao County has only US$109. The implementation of the project will bring direct income for villagers in the community. Villagers from 28 villages in 21 townships of 5 counties will benefit from the project.

(1) Increase the income of community villagers

Villagers participating in the afforestation and management can obtain both labor income and carbon and wood revenue. The 12,745 villagers of 3,231 farmer households will benefit from the project, of which 5,384 are ethnic minorities, accounting for 42.2%. During the project period, a total revenue of US$9.44 million will be generated, of which US$5.02 million will be labor income, US$3.55 million will be the sales of timber and non-wood forest products, plus US$870,000 of CERs sale. The annual net income per capita will increase by US$24.7, about 10.68% of the 2006 level. The extra income is particularly important for ethnic minorities living in the mountains. Beichuan County is the only autonomous county of Qiang Minority in China. The population of Qiang Minority accounts for more than two-thirds of the county's population. 473 people of the Qiang Minority of Beichuan County will be the beneficiaries, accounting for 80% of the total population of the project area.

(2) Broaden the employment channels for community villagers

The project will create 1.105 million short-term employment opportunities for local or surrounding farmers. These job opportunities come from project activities such as planting, tending and thinning. The project will also create 38 long-term job opportunities during the crediting period.

(3) Access to all timber, firewood, and forest by-product revenue

After the forest is natural, villagers can collect firewood and all the forest by-products on the forest land through forest management activities such as tending and thinning; after the project crediting period ends, the forests and wood will be returned to the community villagers.

5.3 *Environment*

(1) Conserve biodiversity and ecosystem integrity

The project area is located in one of the hotspots of global biodiversity. The project sites are all around the nature reserve. Many rare and endangered wild animals (giant pandas, golden monkeys, leopards, etc.) live in the nature reserves. For example, the Li County project site is 8 km away from the Miyaluo nature Reserve in Sichuan Province; the project sites of Luoyigou Village, Weiba Village, Dongqiao Village and Hexi Village in Qingchuan County are 35 km away from the Tangjiahe Nature Reserve, and the project site of Yaolin Village of Dongyang River is only 1 km away from the Dongyanggou Provincial Nature Reserve; the project site of Yongxing Village of Beichuan County is 8.5 km away from Piankou Provincial Nature Reserve, and the project sites of Zhenghe and Anmian Village are 2 km from Xiaozhaigou Provincial Nature Reserve. Most of the project sites are covered with shrubs and herbaceous plants, with poor biodiversity is poor and no rare and endangered species in the plot. The project uses local tree species for vegetation restoration, which will contribute to biodiversity conservation. It will not only provide corridors for wild animals and promote the breeding of wild animals, but also increase the habitat area of threatened species and improve their living environment. At the same time, the project also created a source of income for the local community. This will help to reduce activities such as poaching, firewood collection, illegal logging, and collection of non-wood forest products in the nature reserve by the local community, thus reducing the threats to biodiversity.

(2) Reduce soil erosion

The implementation of the project will directly increase the forest area by 2,251.8 ha, so that the vegetation coverage rate in the project area will be improved and the land degradation will be eased, thus effectively controlling the severe soil erosion in the project community.

(3) Improve regional environment

The implementation of the project adds greening to the barren slope and barren hills. The newly planted forest and vegetation can not only conserve water sources, maintain water and soil, but also regulate local climate, thereby improving local ecological environment, protecting farmland, and alleviating and reducing landslides, droughts, floods, and other natural disasters.

5.4 Policy

The implementation of the project was given high attention by the State Forestry and Grassland Administration, the Forestry Department of Sichuan Province and the county governments and forestry offices involved in the project. The Forestry Department of Sichuan Province established a leading group for forest carbon. The group was led by the head of the Department, with the deputy head as the vice chair. The group is composed of Greening and Afforestation Division, Planning and Finance Division, Foreign Exchange and Cooperation Division, Wildlife Protection Division and Science and Technology Division. There's a forest carbon office under the group, affiliated to the Greening and Afforestation Division. The group has full (part-time) staff, and sets up a forest carbon information center at the Sichuan Forestry Inventory and Planning Institute. Meanwhile, the forestry authorities of project counties also established management offices and identified personnel who are responsible for project organization, coordination, and implementation. The success of the forest carbon project in northwest Sichuan has created a precedent for the forestry sector in the province. It filled the gaps in Sichuan's forestry sector, increased its visibility, and explored a path of success for its forestry to move in line with international standards and make better use of forestry's ecological service. The project brought in new channels for forest farmers to increase their income.

The project regards technical management regulations as the basis for all work. The compilation of the *Technical Document Collection of Investigation and Design of CDM Afforestation and Reforestation Carbon Sink Project* includes the "Investigation and Design Guidance on CDM Afforestation and Reforestation Carbon Sink Project (Trial version)", "Guidelines for the Measurement Methodology of Carbon Baseline of FCCB projects in Sichuan Province" and "CDM A/R Basic Information Investigation Form." These guidelines not only provided the basis for the development of specific projects, but also offered references for developing forest carbon projects in the province in future; the forest carbon database laid down the foundation for future project site selection and project development.

5.4.1 Market

At present, the Dadu River Afforestation Bureau and the Beijing Shanshui Conservation Center signed the cooperation agreement on carbon dioxide emission reduction trade with the Hong Kong Carbon Care Asia for the first verification period (2007–2012). The project can generate about 50,000 tons of CO_2 emission reductions, and is expected to realize a return of approximately US$300,000.

6 Experience and Lessons

6.1 Experience

(1) Strengthen organization and management is the guarantee for the project

From the Forestry Department of Sichuan Province to the project owner Dadu River Afforestation Bureau, from the county and township governments of the project to the project's management authorities, leaders at all levels attach great importance, establish management agencies at each level and formulate management plans to provide organizational guarantees for the forest carbon project. The forest carbon office of the Forestry Department of Sichuan Province makes the project the focus of its work. During the 3-year project period, the office issued 28 editions of *FCCB Project Newsletter* to keep a record of each major event. The newsletters are distributed to relevant leaders and personnel for them to learn and understand project activities, and to pay more attention and highlight forest carbon projects. The Dadu River Afforestation Bureau, as the owner of the project, attaches greater importance to the work of carbon sinks. It held a number of thematic meetings to analyze, study and deploy tasks, identify the leadership and management departments. The leading group of forest carbons is set up, which is headed by the director general. The Party committee secretary/Executive Deputy Director General works as the deputy head and is responsible for carbon sink work. The Afforestation Division is the management body of forest carbon project. A forest carbon office was set up with three full-time staff members who were assigned with specific job responsibilities. A corresponding management plan was formulated and work plans were drawn up to ensure the smooth development of carbon project. The Forestry Bureaus of Li County, Mao County, Qingchuan County, Pingwu County, Beichuan County attached great importance to the project. Leadership groups were set up. The professional and technical personnel were assigned to take charge of project investigation and design. This not only trained the team, but also provided human resource, material and a financial guarantee for smooth progress of the project. In order to ensure the sustainable construction of the project, in June 2010, the "Project Management and Coordination Committee" was set up to coordinate and manage the project.

(2) Organize a specialist team is the basis of the project

CDM A/R project is a—new model of international cooperation. From project development, application, certification to registration, from project implementation to CER issuance of emission reduction and international transaction, its rules and procedures are very complicated, involving extensive technical knowledge and cooperation field. First, the project relies on CI, TNC, Beijing Shanshui Conservation Center, Chinese Academy of Forestry, Sichuan Forestry Inventory and Planning Institute, Sichuan Forestry Science and Research Institute and other organizations and R&D institutions as technical support to promote project development; second, an expert group of 15 persons was set up to provide technical guidance for forest carbon projects in Sichuan Province; third, training, seminars, visits and other forms and methods were used to have in-depth understanding of forest carbon expertise, and actively carry out exchanges in carbon technology and experience, so as to timely understand, obtain and master new technologies, standards, and requirements for carbon sinks; fourth, experts are invited and technical personnel are organized to roll out CDM A/R project in terms of basic knowledge, county-level basic information investigation, village-level basic information collection, carbon sink measurement and monitoring methods, carbon baseline investigation, CCB standards and community assessment methods and training seminars. The technician's ability and business performance are improved, which provides necessary human resources and technical preparation for the orderly development of project development and implementation. It also lays down the technical foundation for carbon sink projects.

(3) Extensive cooperation and exchange is the key to the project

As the project is a new field, during the development and implementation process, participants include the relevant forestry departments of the government, the project county government and forestry bureaus (or the nature reserves), the project township government and forestry offices, and NGOs such as CI, TNC, and Beijing Shanshui Conservation Center, as well as technical supports of Sichuan Forestry Inventory and Planning Institute, Sichuan Forestry Science and Research Institute, Sichuan Agricultural University, the Chengdu Institute of Biology of China Academy of Science, plus the funding support by 3M and the project owner, the Dadu River Afforestation Bureau; the participants are officials, technical experts, company representatives, and community villagers; the work includes basic project information collection, carbon baseline investigation, biodiversity investigation, and community participatory appraisal as well as project afforestation design, environmental impact assessment, additionality analysis, and PDD preparation; it quires not only project application by competent national authority, but also the approval of the project management authority, DOE validation and registration approval by the CDM-EB. Therefore, the project is an interdepartmental and interdisciplinary cooperation project that has both domestic and international levels, with multiple participating units, extensive personnel, and in-depth exchanges. To ensure the successful development of the project and promote its smooth implementation, all participants must communicate extensively and collaborate. To this end, the project organized a series of seminars, trainings, and visits to create conditions for the exchange between counties and between provinces. Based on actual conditions, all people explored and complemented each other. They established sound relationships, enhanced understanding, and friendship, and laid a solid foundation for accomplishing the project.

(4) Mobilize community participation is the core of the project

The community is the main body of the project. Of the 2,251.8 ha, 99.8% belongs to the village collectives or villagers' groups, involving 3,231 farmer households. They are the main force for project implementation. During the development period, through participatory appraisal, villagers can participate in the discussion of determining project plots, selecting tree species and deciding on development approach. This reflected the will of most villagers and improved the scientific feature and feasibility of the project design. The local people's interest is better reflected in the planning and community development. During the implementation period, it mainly relies on the community to organize the workforce to complete soil preparation, planting, tending and other work. Through the community, the management and protection measures and responsibilities were identified.

(5) Adequate Funding is fundamental to a smooth project

Because the implementation of the project involves project development, DOE verification and certification, project application, registration, afforestation implementation, monitoring, technical training, and other processes, plus the difficulty of afforestation, its cost is much higher than the general afforestation project. Therefore, in terms of financial security, first, the project owners and the forestry authorities of the project counties used the successfully developed case to obtain project funding from national and local governments; second, CI, TNC, Beijing Shanshui Conservation Center, 3M, and other NGOs and enterprises offered to fund during the project development period; third, the project owners and the county forestry authorities raised operational funds necessary for project implementation.

6.2 Lessons Learned

This project is the first CDM A/R project in Sichuan, which not only brings new opportunities for forestry innovation, but also brings about many difficulties and challenges for project development and implementation.

The development of the project lasted three years, which made people feel that the project was particularly complicated and that some people became pessimistic. This had a certain influence on the enthusiasm of developing and implementing forest carbon project. This is mainly because that during the preliminary preparation period, nearly half of the time was devoted to the introduction of new concepts and training of investigation and development approach, research and field visits, but these activities also trained talents for developing future projects.

There is a lack of project supervision, and the survival rate and preservation rate failed to keep up, leading to certain impacts and constraints on the realization of the expected benefits of the project. The postponement of project implementation caused by the earthquake also severely affected the timely implementation of the first verification period.

The project design must be carried out in accordance with the principle of "according to local conditions, with suitable lands and trees". The selection of afforestation plots must meet the requirements of the project with respect to land eligibility. The selection of afforestation tree species must be scientific on the basis of respecting the will of the community people. Due to the fact that some plots are located in the dry river valley area, it is difficult to carry out afforestation, and the mortality rate of newly planted seedlings is high, thus the workload of replanting is heavier than other places.

The implementation schedule of the project must give full consideration to the preparation of seedlings and labor force. In some part of the project, due to inadequate seedlings preparation, the actual afforestation tree species in some plots are inconsistent with the design species.

Restoration Project of Small-Scale Reforestation in Tengchong of Yunnan Province

Jian Ma, Caifu Tang and Biao Yang

Summary

Since land use and land use change contribute 25% of global carbon emission, the restoration of vegetation through forestry activities (such as afforestation/reforestation) will promote the absorption of carbon dioxide, which is the main greenhouse gas in the atmosphere. As a result, in the CDM of the Kyoto Protocol, developed countries can obtain carbon emission reduction by supporting afforestation/reforestation activities in developing countries to offset the pressure of their emission reduction.

In light of the above background, on July 25, 2005, Conservation international (hereinafter CI), the TNC, and the SFGA signed the *Agreement on Developing the Project of Forest Restoration for Climate, Community and Biodiversity* in Beijing. The three parties agreed to the goal of conserving biodiversity, reducing global warming, and promoting the development of community. The three parties will carry out a forest multi-benefits project, namely, the FCCB project, in the southwest mountainous region of China, which is one of the key regions of biodiversity conservation.

Through screening, Tengchong was selected as one of the first demonstration sites, and the surrounding areas of Gaoligong Mountain were considered as the pilot of the project. The project was officially launched in 2005. After more than a years' effort, the restoration project of small-scale reforestation in Tengchong was approved by the third-party verification agency (TUV/SUD) in terms of CDM and Climate, Community, and Biodiversity (hereinafter CCB) review in January 2007. It becomes the first forestry carbon sink project meeting the CCB Gold Standard in the world.

J. Ma · C. Tang (✉)
Sichuan Green Carbon Ltd., Chengdu, China
e-mail: 631592345@qq.com

B. Yang
SEE Foundation, Chengdu, China
e-mail: yangb315@163.com

Z. Lu et al. (eds.), *Forest Carbon Practices and Low Carbon Development in China*, https://doi.org/10.1007/978-981-13-7364-0_6

Upon completion, the 467.7 ha of new afforestation will generate 151,971 tons of carbon dioxide tCER within 30 years, and 433 households from 2,108 villagers are expected to benefit from it. The project will contribute to community cohesion and play an important role in biodiversity conservation and prevention of soil erosion.

1 Project Background

Since 2003, the SFGA, CI, and TNC have actively promoted and enhanced cooperation in areas of biodiversity conservation, sustainable forest management, and carbon sinks. The three parties have already and will continue to promote further cooperation in forestry ecological development and global warming mitigation. In order to promote the implementation of the project, according to the "Framework Agreement for Cooperation between the State Forestry and Grassland Administration of the People's Republic of China and the Nature Conservancy of the United States of America" and the "Framework Agreement for Cooperation between the State Forestry and Grassland Administration of the People's Republic of China and the Conservation International," the three parties agreed to carry out FCCB project. The project is based on forest restoration, while taking into account the benefits of climate, community, and biodiversity. The project is funded by 3M, with project funding of US$3 million. On the basis of full consultation among the three parties, the tripartite representatives signed a project agreement in Beijing on July 25, 2005. The tripartite discussions decided to launch two demonstration projects in Yunnan and Sichuan.

The objectives of FCCB project are dedicated to restoring forest vegetation in mountainous regions in southwest China, mitigating global climate change, improving the production and living environment of villagers in local communities, protecting and establishing biodiversity corridors, realizing comprehensive benefits of forests, and providing demonstration for implementing forest restoration and carbon sink transactions.

This project is China's first forestry carbon sink project funded by enterprises, with NGO and government cooperating for development. It is also the first project to use CCB standard for project development in forest carbon project. During the implementation, enterprises, NGOs, and communities cooperated closely to break down previous conditions in which afforestation projects were completely dominated by governments or enterprises. The project successfully completed carbon sink transactions in the voluntary carbon market. The funds provided financial support for project management and community development. It created a brand new condition for China's forestry development.

2 Project History

In November 2004, the preparatory meeting for implementing FCCB project was held in Kunming. The meeting defined the background and expected objectives of the project implementation and the roadmap for future projects. Experts attending the meeting introduced the international situation and development trend of forest carbon trading.

In November 2004, the Forestry Department of Yunnan Province established the "Leading Group for Forest carbon." The leading group members include the head, deputy head, and relevant offices of the Bureau. The leading group consists of an office that is affiliated to the afforestation division of the Bureau.

In January 2005, the project team held a seminar on the selection of forest FCCB projects in Chengdu. The meeting identified project selection criteria, standardized project selection process, and conducted information collection of project sites in Yunnan and Sichuan.

In July 2005, SFGA, CI, and TNC jointly signed the *Agreement on Developing the Project of Forest Restoration for Climate, Community and Biodiversity* to identify the direction of cooperation and conduct field demonstrations in biodiversity hotspots of Yunnan and Sichuan.

In October 2005, the project team signed the "Grant Agreement" and "Memorandum of Cooperation" with the Forestry Department of Yunnan Province and set up a local implementation team.

In November 2005, the project team organized the first field investigation to screen the candidate sites provided by the carbon sink office of Yunnan Provincial. Finally, the surrounding area of the Gaoligong Mountain Nature Reserve in Tengchong was identified as the target area, and boundary survey of the plot is conducted.

In March 2006, the project team conducted project baseline investigation and community investigation training. After the training is completed, information collection and field investigation were also conducted.

In May 2006, the baseline and community investigations at the Tengchong project site were completed and the baseline investigation report was completed.

In July 2006, the project team conducted training seminars on development methods of carbon sink projects, which summarized, discussed, and promoted community investigation, baseline investigation methods, and preliminary work results.

In July 2006, the 3M held a green action media briefing ("Conserving the Beautiful Nature") and a signing ceremony of the cooperation memorandum of "Yunnan Green Environment Development Foundation" in Kunming. The CI and the Forestry Department of Yunnan Province jointly invested seed funding to establish the "Yunnan Green Environment Development Foundation" and carried out the plantation on the 800 mus on the experimental land.

In October 2006, additional information investigation was completed and PDD was drafted.

In December 2006, DOE was invited to carry out project verification.

In January 2007, the project passed verification and became the world's first forestry carbon sink project that had passed the CCB standard gold medal.

In April 2007, the project launched the first phase of afforestation.

In June 2012, the project monitored the certification and issued the first revenue for carbon sink trade to the community.

3 Project Content

3.1 Project Area

The project is located at Tengchong County of Yunnan Province, covering three towns, five villages, and one state-owned forestry farm (Table 1).

The planned area of the project is 467.7 ha. The goal is to build a mixed forest consisting of native species, such as *Taiwania flousiana*, *Betula platyphylla*, Yunnan pine, and *Alnus cremas-togyne*, through 30 years of efforts. 37.6 ha of the project is directly connected with the Gaoligong Mountain Nature Reserve, and 78.2 ha is adjacent to the natural reserve. The project produces 151,971 tons of carbon dioxide tCER (annual average 5,066 tons of carbon dioxide) during the crediting period of 30 years from 2007 to 2036.

Table 1 Areas covered by the project

Towns, villages, and forestry farm	Administrative villages	Natural villages
Qushi town	Pingdi village Daba village	Minzudui, Dazhuyuan, Lichaiba, Sanjia Village, Daijiazhai, Fangjiazhai
Jietou village	Donghua village	Erdaohe, Banpo, Guanjiazhai, Lijiazhai, Wangjiaying, Cunjiaying, Zhaojiapotou,Yangjiazhai, Yujiazhai, Hengzhaizi
	Zhoujiapo village	Mopanshi
Houqiao town	Shangjie village	Haoziba
Sujiang forestry farm		

3.2 *Project Target*

The project implementation will help alleviate poverty and conserve the biodiversity and reduce the soil erosion. The project also contributed to achieve local sustainable development. This is mainly reflected in the following aspects:

Reforestation activities will be carried out in the border zone of natural reserves and buffer zones will be set up between communities and natural reserves to reduce the dependence of community residents on natural reserves.

Inhibiting the growth of *Eupatorium adenophorum*, an invasive alien species, and controlling its expansion into natural reserves; mitigating the threat of invasive alien species to biodiversity conservation.

Controlling soil erosion. The project site is located in the tributary of the Longchuan River in the upper reaches of the Irrawaddy River, an important international river. The reforestation activities can reduce the impact of soil erosion on this important international river.

Creating high-quality forest vegetation will contribute to the mitigation of climate change.

Increasing community income and alleviating poverty pressure.

3.3 *Implementing Body*

As the project implementing body, the Sujiang Forest Farm under the Tengchong Forestry Bureau was established in 1985. It is located in the northwest of Tengchong County. The farm is situated in Houqiao Town, a state-level port in the northwest of Tengchong County and 54 km away from the downtown of county. The climate type in the area is southwest monsoon with distinct wet and dry seasons and abundant rainfall. The farm has a total of five state-owned forests of various sizes. Its boundaries are Dacaopo, Zhonghe Township, Zhaobi Mountain, and Dabanqing to the east; Yingjiang County to the South; Myanmar (20.5 km long of forest borderline) to the west; and Danzha Forest Farm, Guyong Forest Farm, and Guyong Lunma Village to the north. The soil types are mountain yellow, yellow brown, brown, and dark brown, with slightly acidic feature; the main vegetation are *Yunnan pine*, *Quercus acutissima*, hard broadleaf, soft broadleaf, *Taiwania flousiana*, *Gramineae*, ferns, etc.

The total area of the Sujiang Forest Farm is 13,862.0 ha. According to the ownership right, the forest land can be divided into the following: the state-owned forest land area is 12,961.1 ha, accounting for 93.50% of the total; the collective forest land area is 881.1 ha, accounting for 6.36% of the total; and the individual area is 19.8 ha, accounting for 0.14% of the total. Among the land area of the Sujiang Forestry Farm, 13,465.2 ha is used as forestry land, accounting for 97.14% of the total; 396.8 ha are non-forestry land, accounting for 2.86% of the total. The forest coverage rate is 87.29%, the greening rate is 91.75%, and the total standing timber volume is 2,110,320 m^3.

There are 32 cadres and staff working for the farm, including 24 male employees and 8 female employees; 5 retired employees. Among the current employees, there are 2 engineers, 1 assistant engineer, 2 technicians, and 4 accountants; 26 middle-aged and young employees and 6 elderly workers. The farm's economic income mainly depends on timber production. In addition, the farm also planted 4,500 mu of grass and fruit under the forests, which can bring part of income.

3.4 Operating Mode

During the project development stage, technical guidance is provided by expert team organized by TNC and CI. The team is responsible for drafting PDD. The third-party verification of the project is undertaken by CI with financial support from 3M's grants.

In this project, individuals and collective landowners provided land to the project free of charge. The project was managed by Sujiang Forest Farm, which signed land use contracts with landowners. The costs incurred during project operation will not be borne by landowners.

In the future, the project may generate forest carbon benefits and timber revenues, of which 90% of the proceeds from carbon will be returned to Sujiang Forest Farm and 10% to landowners. During the 30-year period, two thinning are planned in the 20th and 27th years. The input for the thinning will be borne by the Sujiang forest farm. Therefore, all the timber proceeds will belong to the farm. The landowners will not share the benefit. Thirty years later, when the project cycle ends, the project owner takes back the land, the forest and timber are collectively returned to landowners, and the project implementing body does not share the benefit.

The project afforestation fund is borne by the company undertaking the project. The project adopts three kinds of capital investment: national investment, corporate self-raising, and bank loans. The landowner does not bear the investment of the initial afforestation funds. In addition, the company will also employ landowners and other community residents to carry out work such as soil preparation, digging, weeding, fertilization, and afforestation. By participating in these jobs, landowners and other community residents will get a certain amount of labor income.

During the operational phase, the company employs some landowners to carry out management work, such as prevention of grazing, fires, pests, and illegal logging. Employed personnel will get remuneration according to the company's employee entitlement.

During the operation, the afforestation design and afforestation effectiveness inspection and monitoring work will be organized by the company. Local forestry bureau and the Yunnan Green Environment Development Foundation will provide technical support, and the cost incurred will be borne by the company.

The certification of emission reduction is also carried out by the company in connection with relevant DOEs. All cost incurred by the certification is borne by the company.

3.5 Project Duration

The project will last for 30 years (2007–2037).

3.6 Land Eligibility

The land eligibility demonstration of this project uses the "procedure for identifying land eligibility of afforestation/reforestation project activities" adopted by the CDM-EB.

(1) Prove that the project site is a no-forest land at the beginning of the project

Land use/coverage investigation: the purpose of current site investigation is to determine the type of land use/coverage of the current project plot, so as to ensure that the plot meets the requirements for afforestation/reforestation land eligibility. The plots selected for the project are farmland, abandoned farmland occupied by the exotic invasive species Eupatorium adenophorum, and fern-covered plots with sparse shrubs. There are scattered grownup trees in some plots, but they cannot meet the Chinese government's criteria for forests.

Land use/coverage mapping: Landsat ETM+ data was used for interpretation on December 30, 2002 to obtain a map of land use/coverage status.

(2) Prove that the project activities are qualified CDM reforestation activities

The results of interview with local villagers on the land use/coverage history show that there have been no reforestation activities since 1989 on selected plots.

The 1989 Landsat TM data are used to interpret land use/coverage results. The result is then compared against the initially selected land plot to determine whether the selected site is no-forest land in 1989.

(3) Baseline investigation

An investigation of natural renewal of the plots was conducted to determine whether the project plots did not have natural renewal or the natural renewal was insufficient to form a forest.

3.7 Baseline Scenario

1. **Demonstration of baseline scenario**

Since the project is applicable to the "Simplified Baseline and Monitoring Methodologies for Small-scale Afforestation and Reforestation Project Activities under the Clean Development Mechanism Implemented on Grasslands or Croplands" (AR-AMS0001/Version 3), the project baseline scenario demonstration is completed according to the methodological guidance.

(1) Land use of the project site

Without the project, due to certain obstacles, the land use of the project site will maintain the status quo and remain degraded. The reason is as follows:

At present, the project land is mainly farmland, abandoned cultivated land covered by *E. adenophorum*, and grassland with small scattered shrubs and high fern coverage. Due to long-term farming and abandonment, the seed layer in the soil has been completely destroyed, and thus natural renewal cannot occur on these lands. A baseline investigation conducted confirmed this conclusion.

For agricultural land plot, a total of 66.2 ha of cultivated land will require reforestation. On one hand, local farmers have enough arable land. On the other hand, the selected arable land is far from the village and the land fertility is low. The results of the community investigation also showed that farmers have no economic returns in terms of cultivation. In particular, after the labor cost is factored, the land has no profit. Therefore, local farmers believe that if there are other land use methods that are economically attractive, they are willing to give up farming on these plots. Through community investigation and interviews with stakeholders, it is known that the only government investment program for the land is a program for conversion of farmland to forest. However, as the quota of the program has gradually declined, this area has been unable to obtain funds from the government for the reforestation. Therefore, without the intervention of the project, the land use will still maintain farmland or become abandoned farmland and continue to degrade.

According to community supplementary investigation, there are sporadic seasonal grazing on some of the project sites. However, because there are no other land use methods that are economically attractive, such as returning farmland to forests and returning grazing land to grassland, these land plots will continue to maintain the current mode of seasonal grazing and will not change. After the project is carried out, the local community will stop grazing on these plots since the project benefit will be greater than the current seasonal grazing; and since there are a large number of grazing lands in the neighboring areas, even if the grazing on some project plots is completely transferred to the adjacent plots, the pressure caused by grazing on these plots is still far below the bearing capacity of grazing on these plots.

Without the implementation of the proposed small-scale reforestation project, investment, technology, and social and market risks will prevent investors or local villagers from converting the land to forest land. All the land (land use/coverage type) plots will remain as it is.

The results of socioeconomic investigation and interviews with stakeholders show that the use of similar land in adjacent areas has not changed. Due to the lack of investment and preferential policy, community residents around Gaoligong Mountain have no incentive to carry out afforestation on sites that are far from the community. The results of field investigation also indicate that for project participants, they currently do not change the type of land use/coverage of cultivated land, abandoned farmland, and grassland intended for the reforestation plots.

The land owned by Sujiang Forest Farm is a seriously degraded land and is located far away, so the economic benefit of reforestation is less attractive. At present, the farm also has no ability to carry out afforestation work on these lands, because as an independent accounting entity, the farm has no way to obtain government's afforestation project; moreover, the annual income of the farm can only meet the staff's salary and the regeneration on the cut-off land. It is impossible to put additional funding in the afforestation of these land plots.

(2) Natural renewability on the project site

The natural renewability of the project land is very limited. In any case, without the intervention of the project, it will never be converted into forest vegetation. This is mainly because of the following:

Long-term farming and cultivation have completely destroyed the seed layer at the bottom of the land block;
Because of the large size of the plots, the seeds in neighboring forests are impossible to be disseminated to form forest coverage;
Some natural renewed trees and trees on the plots along the project boundary can all be seeds. However, through field surveys, it is known that the coverage of these scattered trees does not reach the lower threshold of Chinese forest standards, and seed development and tree seedling growth are prevented from growing by *Eupatorium adenophorums* and ferns that have higher coverage;
The results of field surveys on these plots showed that only a very small amount of seedlings were found in the plots.

2. **Additionality**

The additionality assessment of the project uses the "additionality assessment" method in Appendix B of the approved methodology.

(1) Investment obstruction

Villagers cannot obtain commercial loans from banks for afforestation activities. Agricultural products are the main source of income for the region. The local productivity level is relatively low, with per capita GDP only at US$228 per year, and the lowest being the village of Lisu ethnic group at US$75. Under the situation, local villagers are still below the national poverty line. In addition, the revenue of wood products and non-wood forest products are only available until a considerable period of time after afforestation. In addition, the income of carbon sink trade will also not be available until the project is started for a period of time. Therefore, the lack of necessary funds makes it impossible for villagers to make large investment at the initial stage of afforestation for creating high-quality forests.

Due to the high risk of investment and the uneconomical attractiveness of the land, ordinary reforestation projects cannot obtain commercial loans from banks. However, through the implementation of the forest carbon project, it may obtain commercial loans from banks, and local government is also willing to provide some matching funds.

The Sujiang Forest Farm is an independent accounting economic entity. At present, the resources of the farm can only meet the annual cutting allowance. Each year, in addition to paying wages, the remaining timber income can only be used for renewal work. In addition, due to the low yield of forest farms, it is difficult to obtain commercial loans from banks. Therefore, without the reforestation project, the farm is not able to carry out the reforestation activities on the abovementioned plots.[1]

It is estimated that a total of 171,313 working days of manual input will be required for the project during afforestation and management. Among these, 49,012 working days are required for soil preparation and planting, with each working day needing investment of RMB 20. So a total investment of approximately RMB 1,000,000 is required. The farm will bear all project start-up expenses, including labor costs, of which 30% is from the annual benefit of the farms, and 70% from commercial loans.

(2) Technical obstacles

Lack of seed: Community surveys show that local villagers find it difficult to obtain high-quality seeds;

Villagers lack the knowledge of forest ecology and woodland management techniques.

(3) Obstacles of social conditions

Lack of experienced, well-trained employees: Community surveys show that local villagers lack trainings in supply of seed, afforestation, forest fire prevention, and forest pest management.

Lack of community organization and management: Individual households are too weak in operational investment and marketization, especially when they try to bring wood and non-wood forest products to market, they need more time, energy, and capacity than agricultural products. In addition, there is a lack of organizational structures to circumvent some of the technical barriers mentioned above.

For forest farms, there are not many technical obstacles but the main barriers are investment.

3. **Non-permanence**

The project uses tCER to solve non-permanence issues. It is estimated that during the crediting period (June 1, 2007–May 31, 2037), a total of 152,000 tons of carbon could be generated.

---

[1]China states that the poverty line for remote areas and areas inhabited by minority ethnic groups is less than RMB 2,700 (US$370) per capita GDP (National People's Congress, 1996, catalogue of national poverty county).

3.8 Estimate of Net Anthropogenic GHG Removals by Sinks

1. **Estimate of actual GHG removals by sinks**

The carbon stock at the beginning of the project is the same as the predicted net baseline GHG removals by sinks. In the remaining years, the carbon stock within the project boundary at time t can be calculated using the following formula:

$$N_{(t)} = \sum_{i} (N_{A(t)i} + N_{B(t)i}) \cdot A_t$$

$$N_{A(t)} = T_{(t)} \cdot 0.5$$

$$N_{B(t)} = T_{(t)} \cdot R \cdot 0.5$$

$$N_{(t)} = SV_{(t)} BEF \cdot WD$$

Among them, $N_{(t)}$, under the project scenario, is the total biomass carbon pool at time *t*, tons of carbon/ha; $NA_{(t)i}$, under the project scenario, is the carbon pool aboveground for carbon layer *i* at time *t*, tons of carbon/ha; $NB_{(t)i}$, under the project scenario, is the belowground carbon pool in the carbon layer *i* at time *t*, ton of carbon/ha; A_i, *i* is the area of carbon layer, hectare; $T_{(t)}$, under the project scenario, is the aboveground biomass at time *t*, tons of dry matter per hectare; R is the root and stem ratio; 0.5 is the carbon content of dry matter, tons of carbon/ton of dry matter; $SV_{(t)}$ is the living standing stock under project scenario, cubic meters/hectare; WD is the standard wood density, tons of dry matter per cubic meter; BEF is the biomass expansion factor; and *i* is the carbon layer *i*.

According to the seven sample plots of forest inventory conducted by the Yunnan Provincial Forestry Investigation Planning Institute in 1992, 1997, and 2002, the biomass equations for the intended tree species of the project are as follows:

Taiwania flousiana: $SV_{(t)} = NT_{(t)} \cdot 2.251370 \cdot (1-e^{-0.0452113A})^{5.942134}$
Yunnan pine: $SV_{(t)} = NT_{(t)}0.200346 \cdot (1-e^{-0.054764A})^{2.978562}$
Alnus cremas-togyne and *Betula platyphylla*: $SV(t) = NT(t) \cdot 0.880452 \cdot (1-e^{-0.039517A})^{2.912148}$.

Among these, $SV_{(t)}$ is the amount of live standing stock under the project scenario, m^3/ha; $NT_{(t)}$ is the quantity, plant/ha; A is the tree age, year.

The forecast results are shown in Table 2.

2. **Estimate of baseline carbon stock**

There are some scattered wood growing on the project site. Even in the absence of project activities, the natural growth of trees will increase the carbon pool of the scattered wood over time.

For the abandoned farmland that is covered by *E. adenophorum* and grassland with high fern coverage, since the biomass does not increase, it can be assumed that the carbon stock of these two types of lands is subject to zero change.

Table 2 Estimate of carbon stock for the afforestation project during 2007–2037

Year	Tons of carbon/aboveground biomass	Tons of carbon/belowground biomass	Total/tons of carbon
2007	−742	−183	−925
2008	37	8	45
2009	231	51	283
2010	575	129	703
2011	1029	232	1261
2012	1565	355	1920
2013	2164	495	2659
2014	2813	649	3462
2015	3503	816	4319
2016	4227	993	5220
2017	4978	1182	6159
2018	5752	1380	7131
2019	6543	1586	8129
2020	7347	1799	9146
2021	8158	2018	10,176
2022	8969	2241	11,210
2023	−7620	−1702	−9322
2024	6252	1505	7757
2025	8134	2080	10,214
2026	8667	2234	10,901
2027	9178	2384	11,563
2028	9667	2529	12,196
2029	10,129	2668	12,796
2030	−18,408	−4489	−22,898
2031	5861	1459	7320
2032	8115	2165	10,280
2033	8346	2238	10,584
2034	8551	2304	10,855
2035	8732	2363	11,095
2036	8887	2415	11,302
2037	9017	2460	11,476
Total	141,397	36,546	177,943

Table 3 Baseline carbon layer

Name of carbon layer	Number of plot	Area (ha)	Scattered wood			
			Number of trees per hectare	Average BDH (cm)	Average height (m)	Average life (year)
Cultivated land, with small amount of scattered wood and *Eupatorium adenophorum* at the edge of the plot	5	66.2	40	12.4	6.5	10
High coverage of grassland and fern, small amount of scattered wood and small shrubs	5	351.9	29	14.9	6.1	15
Abandoned farmland, mainly covered by *Eupatorium adenophorum*, no scattered wood	3	49.6	–	–	–	–

In order to accurately estimate the net baseline removals by sinks, the proposed project divides the land into corresponding carbon layers (Table 3).

The main tree species of scattered wood are *Alnus cremas-togyne*, *Betula platyphylla*, *Juglans sigillata*, and *Acer amplum*.

The amount of scattered wood growth is calculated using the following formula in the CDM Monitoring Methodology for Small-scale Reforestation Project Activities that has been approved:

$$B_{(t)} = \sum_{i}^{I} (B_{A(t)i} + B_{B(t)i}) \cdot A_t$$

Among these, $B_{(t)}$, in the absence of project, is the baseline biomass carbon pool, tons of carbon; $B_{A(t)i}$, in the absence of project, is the aboveground biomass carbon stock of carbon layer *i*, tons of carbon/hectare $B_{A(t)i}$, in the absence of project, is belowground biomass carbon stock of carbon layer *i*, tons of carbon per hectare; Ai is the area of carbon layer *i*, ha; *i* is the number of carbon layers; and *I* is the total number of carbon layers.

For aboveground biomass, $B_{A(t)}$ is calculated for each carbon layer and is based on the following formula:

$$B_{A(t)} = M_{(t)} \cdot 0.5$$

Among these, $B_{A(t)}$, in the absence of the project, is the aboveground biomass carbon stock of the carbon pool at time *t*, tons of carbon/ha; $M_{(t)}$, in the absence of the project, is the aboveground biomass at the time *t*, tons of dry matter/hectare; and 0.5 is the conversion coefficient for dry matter and carbon, tons of carbon/ton of dry matter.

According to the approved baseline methodology and monitoring methodology for small-scale reforestation projects, in the absence of project, there is no significant change in aboveground biomass or belowground herbaceous biomass of perennial living wood plants, or the biomass is reduced, the baseline carbon stock change is assumed to be zero, and the baseline carbon stock of the carbon pool remains the same as that of all carbon stock measured at the start of the project. Otherwise, in the absence of project, the net baseline greenhouse gas removal must be equal to the changes in the carbon pool of perennial wood living plants or herbaceous belowground carbon pool, for example,

$$M_{(t=0)} = M_{grass} + M_{woody(t=0)}$$

If

$$M_{woody(t=n-1)} + g \cdot t < M_{woody-max}$$

Then

$$M_{(t=n)} = M_{grass} + M_{woody(t=n-1)} + g \cdot t$$

If

$$M_{woody(t=n-1)} + g \cdot t \geq M_{woody-max}$$

Then

$$M_{(t=n)} = M_{grass} + M_{woody(t=n-1)} + g \cdot t$$

Among these, $M_{(t)}$, in the absence of a project, is the aboveground biomass at time *t*, tons of dry matter per hectare; M_{grass}, in the absence of a project, is the aboveground herbaceous biomass at time *t*, tons of dry matter per hectare; $M_{woody(t)}$, in the absence of a project, is the biomass of perennial wood at time *t*, tons of dry matter per hectare; $M_{woody\text{-}max}$, in the absence of a project, is the maximum aboveground biomass of perennial wood at time *t*, tons of dry matter per hectare; *g* is the annual growth rate of perennial wood biomass, tons of dry matter/(hectare litre); *t* is the time increment, here it is 1 year; and *n* is the number of iterations, which refers to the number of years after the project is launched, year.

For belowground biomass, $B_{B(t)}$ is calculated for each carbon layer *i* according to the following formula:

$$B_{B(t=0)} = 0.5 \cdot \left(M_{grass} \cdot R_{grass} + M_{woody(t=0)} \cdot R_{woody}\right)$$

If

$$M_{woody(t=n-1)} + g \cdot t < M_{woody-max}$$

Then

$$B_{B(t=n)} = 0.5 \cdot \left[M_{grass} \cdot R_{grass} + \left(M_{woody(t=n-1)} + g \cdot t\right) \cdot R_{woody}\right]$$

If

$$M_{woody(t=n-1)} + g \cdot t \geq M_{woody-max}$$

Then

$$B_{B(t=n)} = 0.5 \cdot \left(M_{grass} \cdot R_{grass} + M_{woody-max} \cdot R_{woody}\right)$$

Among them, $B_{B(t)}$, in the absence of a project, is the belowground biomass carbon stock at time *t*, tons of carbon/hectare; M_{grass}, in the absence of a project, is the aboveground biomass of herbaceous vegetation at time *t*, tons of dry matter/hectare; $M_{woody(t)}$, in the absence of a project, is the perennial wood biomass at time *t*, ton of dry matter/hectare; $M_{woody\text{-}max}$, in the absence of a project, is the maximum perennial wood biomass at time *t*, tons of dry matter/hectare; R_{woody} is the wood root and stem ratio, tons of dry matter/ton of dry matter; R_{grass} is the herbaceous root and stem ratio, tons of dry matter/ton of dry matter; *g* is the annual wood biomass growth, tons of dry matter (hectare year); *t* is the time increment, here it is 1 year; *n* is the number of iterations, which refers to the number of years after the project is launched, year;

and 0.5 is the conversion coefficient of dry matter and carbon, tons of carbon/ton of dry matter.[2]

In the project land plot, there are scattered wood on the edge of cultivated land and the fern-covered grasslands, and these scattered wood will increase in carbon stock due to continued growth. According to the IPCC LULUCF Good Practice Guide, due to changes in management practices, the biomass of herbaceous vegetation will change in the first few years (for example in 20 years). After that, if there is no change in management, the herbaceous vegetation biomass will remain constant or only slightly change. The agricultural land and grassland involved in this project have been formed at least in the 1980s (20 years ago). Therefore, it can be assumed that the biomass of fern-covered plots and abandoned farmland covered by *E. adenophorum* remains constant. The scattered shrubs that grow on the edge of fern-covered grassland and cultivated land also remain the same status for many years (since deforestation in the 1950s), so they can also be regarded as constant. It can be seen that there is no increase in grassland's belowground biomass and biomass of scattered shrubs, such as M_{grass} and M_{woody}. Therefore, for baseline scenario, we only consider the growth of existing scattered wood, and the biomass changes of herbs and shrubs are negligible.

Table 3 lists the area of each baseline carbon layer, the average number of trees per hectare, the age of trees, and DBH per hectare in each carbon layer. Most of the scattered wood is *Alnus cremas-togyne* and *Betula platyphylla*, under natural growth conditions, and the age of maturity is about 30 years. According to the analysis of seven sample plots conducted by the Yunnan Institute of Forestry Inventory and Planning in 1992, 1997, and 2002, under natural conditions, the growth of living standing trees of *Alnus cremas-togyne* and *Betula platyphylla* can reach 0.521457 cubic meters/strain in 50 years (standard deviation 0.10642 cubic meters/strain). Using the average wood density and biomass expansion factor of *Alnus cremas-togyne* and *Betula platyphylla*, it was estimated that the aboveground carbon reserve in the carbon pool is approximately 0.537423 tons of dry matter per strain for 50 years. This value can be set as the maximum aboveground biomass per tree ($M_{woody\text{-}max}$), and the average annual growth of each tree is 0.010748 tons of dry matter. By calculation, the aboveground biomass of scattered wood in farmland is 0.04252 tons of dry matter per tree, and the aboveground biomass of scattered trees in grassland is 0.058219 tons of dry matter per tree. This means that over the 30-year crediting period, the scattered aboveground wood biomass cannot reach the maximum biomass. For conservative calculation, all biomass expansion factors use 1.5. Based on the number of scattered wood plants listed in Table 3, the average scattered wood biomass around the farmland is estimated to be 0.43 tons dry matter/hectare, and the average biomass growth rate of scattered wood in the grassland is 0.31 tons dry matter/hectare. The survey and summary results are shown in Table 4. Non-wood vegetation biomass is set as a constant.

---

[2]IPCC. Good Practice Guidance for Land Use, Landuse Changeand Forestry. 2003. http://www.ipccnggip.iges.or.jp.

Table 4 Biomass carbon stock at the beginning of project

	Aboveground biomass carbon reserve	Belowground biomass carbon reserve	Total
Farmland	1.51	1.03	2.54
Grassland	2.30	8.33	10.63
Abandoned and deserted land	3.50	1.60	5.10

Unit Tons of carbon/hectare

See Table 5 for the biomass of scattered wood before the project start.[3]

3. **Leakage estimate**

The major sources of potential leakage in the project include the following: The transfer of pastoralism and new deforestation activities resulted from agricultural activities after afforestation; N_2O emissions resulted from the application of certain compound fertilizers during afforestation; and the machinery used in the project, for example, transportation of seedlings and patrols.

(1) Transfer of activities during the project

Local villagers have enough agricultural land, so the transfer of agricultural land and the transfer of farmers after the project are hard to take place. There are scattered grazing behaviors in 239.5 ha of the Haoziba community in this project. According to the community survey, there are 24 buffaloes, 56 cattle, and 148 sheep in the Haoziba community, and the grazing area is 530 ha. As the land selected for the project is far away from the community, there are usually only three buffaloes, two cattle, and 12 sheep on the project site. Even if all livestock were transferred to the rest of the grazing land, the grazing density was only 0.1 buffalo, 0.2 cattle, and 0.5 sheep per hectare, which is lower than the IPCC's default value of 1 cattle and 4.9 sheep per hectare on average. It is also far below the reference value for grazing load of Tengchong County (0.9 cattle, 0.7 buffalo, and 4.5 sheep per hectare). In summary, the transfer of grazing can be counted as 0.

The leakage was counted as 0 due to the fact that the project activity transfer is less than 10%.

(2) Leakage caused by nitrogen-containing compound fertilizer during afforestation

A compound fertilizer containing approximately 10% nitrogen will be used during forest management. Among these, the fertilization is 50 g per tree for the first year of afforestation, 50 g in the second year, 100 g in the third, and 100 g in the fourth. In order to verify that the leakage caused by N_2O emissions is less than 10% of the anthropogenic removal by sinks, the N_2O direct emission is calculated using the default formula provided by IPCC GPG AFOLU:

[3]National Forestry and Grassland Administration, Forestry Investigation Manual, 1994.

Table 5 Estimate of net baseline GHG removals by sinks

Year	Baseline carbon reserve/tons of carbon			Net baseline GHG removals by sinks/(tons of CO_2 year^{-1})		
	Aboveground biomass	Belowground biomass	Total	Aboveground biomass	Belowground biomass	Total
2007	1285.4	3128.8	4414.2	253.3	62.4	315.7
2008	1354.5	3145.8	4500.3	253.3	62.4	315.7
2009	1423.5	3162.8	4586.4	253.3	62.4	315.7
2010	1492.6	3179.9	4672.5	253.3	62.4	315.7
2011	1561.7	3196.9	4758.6	253.3	62.4	315.7
2012	1630.0	3213.9	4844.7	253.3	62.4	315.7
2013	1699.8	3230.9	4930.8	253.3	62.4	315.7
2014	1768.9	3248.0	5016.9	253.3	62.4	315.7
2015	1838.0	3265.0	5103.0	253.3	62.4	315.7
2016	1907.1	3282.0	5189.1	253.3	62.4	315.7
2017	1976.1	3299.0	5275.2	253.3	62.4	315.7
2018	2045.2	3316.1	5361.3	253.3	62.4	315.7
2019	2114.3	3333.1	5447.4	253.3	62.4	315.7
2020	2183.4	3350.1	5533.5	253.3	62.4	315.7
2021	2252.4	3367.2	5619.6	253.3	62.4	315.7
2022	2321.5	3384.2	5705.7	253.3	62.4	315.7
2023	2390.6	3401.2	5791.8	253.3	62.4	315.7
2024	2459.7	3418.2	5877.9	253.3	62.4	315.7
2025	2528.7	3435.3	5964.0	253.3	62.4	315.7
2026	2597.8	3452.3	6050.1	253.3	62.4	315.7
2027	2666.9	3469.3	6136.2	253.3	62.4	315.7
2028	2736.0	3486.3	6222.3	253.3	62.4	315.7
2029	2805.0	3503.4	6308.4	253.3	62.4	315.7
2030	2874.1	3520.4	6394.5	253.3	62.4	315.7
2031	2943.2	3537.4	6480.6	253.3	62.4	315.7
2032	3012.3	3554.5	6566.7	253.3	62.4	315.7
2033	3081.3	3571.5	6652.8	253.3	62.4	315.7
2034	3150.4	3588.5	6738.9	253.3	62.4	315.7
2035	3219.5	3605.5	6825.0	253.3	62.4	315.7
2036	3288.6	3622.6	6911.1	253.3	62.4	315.7
2037	3357.6	3639.6	6997.2	253.3	62.4	315.7

$$N_2O_{direct-N_{fertilizer}} = F_{SN} \cdot EF_1 \cdot (44/28) \cdot 310$$
$$F_{SN} = N_{SN-Fert} \cdot (1 - Frac_{GASF})$$

Among these, $N_2O_{direct\text{-}Nfertilizer}$ is the direct N_2O emission due to the use of nitrogen fertilizer in the project boundary, tons of carbon/year; F_{SN} is the N in the compound fertilizer that equals the amount of NH_3 and NO_X, tons of nitrogen/year; $N_{SN\text{-}Fert}$ is the nitrogen usage in composite fertilizer, tons of nitrogen/year; EF_1 is the emission factor, tons of N_2O–N (tons N input)$^{-1}$; $Frac_{GASF}$ is the compound fertilizer that becomes NH_3 and NO_X coefficients, dimensionless; 44/28 is the relative molecular mass ratio of N_2O and N, dimensionless; and 310 is the N_2O global warming potential (effective during the first commitment period).

Since China does not have nationally uniform parameters, the value of EF_1 (1.00%) adopts the default values of NO_X and NH_3 released from compound fertilizers and organic fertilizers (0.1 and 0.2, respectively) in the IPCC 2006 GPG AFOLU reference manual.

The total amount of direct N_2O emission generated in the project is estimated at 222 tons of carbon, which is negligible.

In the project area, mid-sized diesel agricultural vehicles are generally used. In view of the seedlings needed for the project, and the need for future thinning and commercial harvesting, as well as the need for fertilizer transport, the leakage generated during the use of machinery needs to be estimated. For mid-sized agricultural vehicles, the diesel emission factor is 2.6353 kg/L diesel (China's national initial information). Therefore, after calculation, the project's leakage due to mechanical use during the crediting period is 9.15 tons of carbon, which is also very small and negligible.

4. **Estimate of net anthropogenic GHG removals by sinks**

Net anthropogenic GHG removals by sinks = actual net GHG removals by sinks—baseline GHG removals by sinks—leakage.

The results of net anthropogenic GHG removals by sinks are shown in Table 6.[4]

3.9 Main Tree Species and Modes of Afforestation

The project is designed based on the following standards:

Technical Regulations for afforestation (GB/T15776–1995);
Non-commercial Forest Construction—Guide Principle (GB/T18337.1–2001); Non-commercial Forest Construction—Regulation of Plan and Design (GB/T18337.2–2001); Non-commercial Forest Construction—Technical Regulation (GB/T18337.3–2001);

[4]IPCC. 2006. IPCC Guidelines for National Greenhouse Gas Inventories. (2006) http://www.ipcc-ng-gip.iges.or.jp.

Table 6 Estimate of net anthropogenic GHG removals by sinks

Year	Baseline net GHG removals by sinks (tons of carbon year^{-1})	Actual net GHG removals by sinks (tons of carbon year^{-1})	Leakage (tons of carbon year^{-1})	Net anthropogenic GHG removals by sinks (tons of carbon year^{-1})
2007	315.7	−16,185	0	−16,501
2008	315.7	45	0	−270
2009	315.7	283	0	−33
2010	315.7	703	0	388
2011	315.7	1261	0	945
2012	315.7	1920	0	1604
2013	315.7	2659	0	2344
2014	315.7	3462	0	3147
2015	315.7	4319	0	4003
2016	315.7	5220	0	4904
2017	315.7	6159	0	5844
2018	315.7	7131	0	6815
2019	315.7	8129	0	7813
2020	315.7	9146	0	8830
2021	315.7	10,176	0	9860
2022	315.7	11,210	0	10,895
2023	315.7	−9322	0	−9637
2024	315.7	7757	0	7441
2025	315.7	10,214	0	9899
2026	315.7	10,901	0	10,585
2027	315.7	11,563	0	11,247
2028	315.7	12,196	0	11,880
2029	315.7	12,796	0	12,481
2030	315.7	−22,898	0	−23,213
2031	315.7	7320	0	7005
2032	315.7	10,280	0	9964
2033	315.7	10,584	0	10,268
2034	315.7	10,855	0	10,540
2035	315.7	11,095	0	10,779
2036	315.7	11,302	10,986	
2037	315.7	11,476	0	11,160
Total	9787	161,758	0	151,971

Table 7 Tree species mixed mode, afforestation mode

Tree species and mixed mode	Mixed mode code	Planting percentage	2007 afforestation area (ha)	Total area (ha)
Betula platyphylla + *Alnus cremas-togyne*	M1	1:1	7.2	7.2
Taiwania flousiana + *Betula platyphylla*	M2	1:1	282.6	282.6
Taiwania flousiana + *Alnus cremas-togyne* + *Betula platyphylla*	M3	1:1:1	60.7	60.7
Alnus cremas-togyne	M4	1	4.8	4.8
Alnus cremas-togyne + *Yunnan pine*	M5	1:1	112.4	112.4
Total			467.7	467.7

Design Code for Afforestation Operation (LY/T1607–2003);
Regulations for Forest Tending (GB/T15781–1995);
Tree Seedling Quality Grading of Major Species for Afforestation (GB6000–1999);
Technical Regulations for Cultivation of Tree Seedlings (GB/T6001–1985);
Container Nursery Technology (LY1000–1991).

1. **Soil preparation**

In order to avoid soil erosion, reduce greenhouse gas emissions, and protect existing carbon pool, the project will ban controlled mountain fire and complete soil preparation. The existing non-wood vegetation will be removed in triangle shape along the contour lines in a width of 60 cm, while burrowing site will be cleared on the removal belt. There are two sizes of soil preparation, of which size 40 cm × 40 cm × 40 cm (length × width × height) is used for afforestation of *Taiwania flousiana* and *Yunnan pine*, while the size of 30 cm × 30 cm × 30 cm is for *Alnus cremas-togyne* and *Betula platyphylla*.

2. **Tree species and afforestation model**

In order to reduce the risk (fire, pests, etc.) and improve environmental and social economic benefits, the afforestation model used in the project is mixed block mode. Table 7 lists tree species and afforestation models.

3. **Seedling and nursery**

All the seedlings involved in the project are prioritized by locally cultivated species. If the local seedlings are not enough, they will be purchased from neighboring private seedling companies in Qushi Township, Zhonghe Township, and Tengyue Town, in Tengchong County. Source seeds originate from or collected from local seed orchards, without any clones source.

Seeds of *Alnus cremas-togyne* and *Betula platyphylla* will be collected from the Linjiapu Nursery or the Gaoligong Mountain Nature Reserve; the seed of the *Taiwania flousiana* will be collected from local nursery or existing mother trees. The collection sites include Qushi Township, Jietou Township, Houqiao Town, Beihai Township, and Heshun Town of Tengchong County.

The collected seeds of *Alnus cremas-togyne* and *Betula platyphylla* were planted in plastic bags (5 cm in diameter and 15 cm in height) and compacted with 20% organic soil. This technology helps to increase the initial growth quality of tree species and can effectively increase the survival rate.

4. **Reforestation**

Reforestation activities began in 2007 with a planting density of 2 m × 3 m.

To ensure a higher survival rate and better growth in the early stages of reforestation, weeding is done twice a year for the first 35 years (April–May and September–October). One month after afforestation, survival rate will be checked and supplement will be made as needed.

To ensure the growth of the planted trees, compound fertilizers (nitrogen content 10%) will be used. Among them, for the first year, each tree will be administered 50 g of fertilization, followed by the second year 50 g, the third year 100 g, and the fourth year 100 g.

5. **Forest Management**

Two thinning will be carried out for *Alnus cremas-togyne*, *Betula platyphylla,* and *Taiwania flousiana*. The first will be carried out in the 17th year after afforestation and the second in the 24th year after afforestation. Thinning is not needed for Yunnan pine, but commercial logging can be arranged in the 30th year. The thinning intensity is carried out according to 30% of the volume amount. After the thinning of *Betula platyphylla* and *Alnus cremas-togyne*, due to natural regeneration, no replanting will be carried out, and other tree species will be replanted according to the thinning intensity and actual needs.

3.10 Investment Summarization

The total investment of the project is RMB 6.0203 million, of which machinery and equipment costs RMB 30,900, forest land clearing and basic facilities RMB 30,900, fertilization and weeding RMB 2.4989 million, soil preparation and seedling planting RMB 1.2127 million, and other RMB 349,500.

The source of project funds consists of three channels: bank loans, local government support, and self-raised fund from project participants. The proportion and amount of investment are as follows: bank loans (or foundation borrowings) RMB 4.2143 million, accounting for 70% of the total; Forestry Department supporting RMB 602,000, accounting for 10% of the total; local support RMB 602,000, accounting for 10% of the total; and the company has self-raised amount of RMB 602,000, accounting for 10% of the total.

3.11 Key Elements for Monitoring

1. **Baseline net GHG removals by sinks**

Based on the approved small-scale reforestation methodology, the baseline net GHG removals by sinks do not need to be monitored.

2. **Actual net GHG removals by sinks**

Project participant will determine the change in carbon stock by measuring and monitoring the reforestation area of the project. While monitoring project boundary, random sampling is also conducted for each carbon layer. The sampling is done in accordance with Section 4.3.3.4 of IPCC GPG LULUCF and throughout the whole credit period. If unusual carbon reserve is detected in certain area of the same carbon layer during the monitoring process, the area will be monitored as a separate carbon layer.

The project boundary is monitored by remote sensing satellite or GPH real-time coordinates. The changes in project boundary will be considered in the calculation of net GHG removals by sinks. The proposed project references Section 4.3.3.2 of IPCC GPG LULUCF to separate the monitoring of each layer using climate, landscape, soil condition, tree species, planting density, tree species combination, etc. By setting fixed sample for monitoring aboveground and belowground biomass, the sample quantity of carbon layers is obtained with ±5% accuracy of 95% confidence interval. When setting permanent monitoring samples, relevant GPS coordinates must also be recorded.

The estimate of carbon layer stock uses random sampling points of layers, and applies the following formula:

$$P_{(t)} = \sum_i \sum_j \sum_k (P_{A(t)i,j,k} + P_{B(t)i,j,k}) \cdot A_{i,j,k}$$

Among them, $P_{(t)}$ is the total carbon stock of the project activity obtained at the time *t* within the project boundary, tons of carbon; $P_{A(t)i,j,k}$ is the carbon layer *i* per unit area of the project activity during the monitoring period, species *j*, age *k*, carbon stock of aboveground biomass at time *t*, tons of carbon per hectare; $P_{B(t)i,j,k}$ is the carbon layer *i* per unit area of the project activity during the monitoring period, tree species *j*, age *k*, carbon stock of aboveground biomass at time *t*, tons of carbon per hectare; and $A_{i,j,k}$ is the carbon layer *i*, tree species *j*, area of age *k*, hectare.

The project plans to adopt a mixed block approach for reforestation. For the combination of different species, sub-carbon layer will be divided. Therefore, sample monitoring is actually monitoring the pure forest model instead of mixed model. The carbon stock is estimated according to the following method:

(1) Project carbon layer division

On the basis of carbon layer, sub-layers are divided according to the years of afforestation, and hierarchical maps are created using GIS. After the first monitoring, based on the possible changes of the project boundary and the planting time of the project, the carbon stock is calculated. If it is greater than or less than the expected variability, the stratification will be further identified. Further layers need to consider the following elements:

Monitoring data of afforestation and project boundary, such as project boundary, the year of afforestation, and actual time.

Forest management monitoring data, such as thinning and fertilization.

Changes in carbon stock in each carbon layer or sub-layer after the first monitoring. If there are similar carbon stock, changes in carbon stock and space, these carbon layers or sub-layers can be combined.

(2) Sampling design

Through the establishment of permanent monitoring plots, measurement, and monitoring of changes in the associated carbon pool of each carbon layer, the rest of the plot should also be treated equally, for example, soil preparation methods, fertilization, harvesting, etc., so as to prevent deforestation during the project crediting period.

In order to accurately obtain the growth status of forest trees, rectangular plots or concentric plots are used. Considering that the project may have thinning and replanting, which might lead to certain distribution of tree classes, a nesting plot method is adopted to complete the sampling.

Nesting plots are the combination of nested big and small sample plots. This is economical for forests with large variations in tree diameters or where the tree diameters and the degree of standing trees are significant (such as natural forest management and forest protection). Monitoring of plantation forest can be done with a single radius circle (or rectangle).

Determining the number of plots: The accuracy of monitoring requires $\pm 10\%$ of the average 95% confidence interval. In the proposed small-scale reforestation project, the number of plots is estimated using the following formula (Avery and Burkhmrt 1994):

$$n = \frac{(\sum_{h-1}^{L} N_h \times s_h)^2}{\frac{N^2 \times E^2}{t^2} + (\sum_{h-1}^{L} N_h \times s_h^2)}$$

Among them, E is the allowable error or half of the pre-set confidence interval, which is obtained by multiplying the average carbon stock by the expected accuracy, for example, the average carbon stock × 0.1 (10% accuracy) or 0.2 (20% accuracy); *t* is the *t* value (95% confidence interval), in fact, *t* is usually set to 2 when the number of samples is unknown; the number of sampling units of N_h carbon layer, N_h = area of carbon layer/sample area; N is the sampling units (all layers); and s_h is the standard deviation of layers.

Random sample plots: The sample plots were set up using the GIS extension module, and the coordinates of sample plots were permanently monitored by GPS records. The sample plot size is 400 square meters (20 m × 20 m) or a sample circle with a radius of 20 m. Reset measurement is performed at each monitoring period.

(3) Monitoring frequency

The project began afforestation in 2007 and set permanent plots for monitoring once every 5 years, for example, 2012, 2017, 2022, 2027, 2032, and 2037, which is the last year of the credit period.

3. **Leakage caused by project activity**

Regarding the leakage from activities used in the project or described in the approved small-scale afforestation/reforestation methodology, it is likely to be monitored during each monitoring period. The proportion of peasant affected by the project activity and the output of major agricultural products shall be measured. However, the percentages of affected households and production output in the project are all below 10%, as a result,[5]

$$L_{(t)} = 0$$

If the monitoring result of the project boundary shows that the percentage of actual cultivated land is greater than 10%, a sample survey must be conducted on the percentage of farmers involved and the output. If the result of the sampling shows that the ratio is greater than 10% and less than or equal to 50%, leakage must be considered. The project leakage calculation formula is as follows:

$$L_{(t)} = P_{(t)} \cdot 0.15$$

[5]Avery T. E. and H. E. Burkhmrt (eds.). 1994. *Forest Measurements*, 4th edition. McGraw-Hill, New York.

At the same time, it is also necessary to monitor the emission of greenhouse gases generated during the implementation of the project, such as the use of vehicles and machinery, as well as fertilizers.

4. **Monitoring of biodiversity**

The project adopted the same methodology as that in "the forest carbon, community and biodiversity project in Southwest Sichuan" for biodiversity monitoring.

5. **Community benefit monitoring**

Since the project has only a few participating family households, and the methodology is similar to "the forestry carbon, community and biodiversity project in Southwest Sichuan)," it is possible to adopt similar method for monitoring community benefit.

4 Project Implementation

4.1 Main Activities

The project implementation is mainly divided into the following three periods:

1. **Project Planning and Development Period (2005–2006)**

(1) Determination of the general location of project sites

The following rules are formulated with respect to the project: convenient traffic; being representative in terms of landform, vegetation, type of plantation, socioeconomic conditions, etc.; the proposed site should be a biodiversity hotspot interesting to relevant international organizations, or next to natural reserves; ecologically fragile zones where biodiversity is relatively enriched, or areas of great significance for the conservation of biodiversity; areas where basic data are relatively complete and have a certain research basis; areas with relatively large areas of plantation forests and have certain scale of afforestation land; and areas with project capacity (initiative, technical capacity), including benefit monitoring conditions, preferably in areas that have undertaken international cooperation projects.

Projects can be combined with others being implemented in the country, taking into account as much as possible the pilot counties in the logging of plantation commodity forests. The final determined site of the project site is located in the neighborhood of Gaoligong Mountain Nature Reserve in Tengchong.

(2) Survey of project land and investigation of community participation willingness

According to the determined location, remote sensing data is used for preliminary determination of the plots, which is in connection with stakeholders interviews to determine the willingness of the community to participate. Through the comparison of forest maps, remote sensing satellite images, and field surveys, the land eligibility was determined and the boundary coordinates of the land were obtained.

(3) Collection of basic project data

The basic data of the project is used to describe the basic situation of the plot before the project activity. Generally, the basic data include topography, landscape, natural vegetation, hydrology, biodiversity, community social economy, etc. The information, after expert analysis, forms a report and enriches the relevant content of the PDD.

(4) Project baseline investigation

According to the plot condition, CDM A/R methodology that meets the criteria is determined. In this project, CDM A/R small-scale afforestation baselines and monitoring methodologies were selected because the project meets the criteria for small-scale projects. According to the methodological content, the baseline investigation content was developed and the baseline investigation fieldwork manual was developed, and relevant field investigators were trained. In order to reduce the workload of baseline investigations, facilitate field work, and improve the correctness and accuracy of carbon measurement and estimation, it is necessary to divide the project area into several homogeneous sample units.

In general, stratification can reduce measurement and monitoring costs. It can also reduce the required sampling effort while maintaining the same level of confidence. This is because, compared to the entire project, the carbon stock changes in each layer are smaller. Stratification can be determined by satellite imagery, aerial photographs, vegetation, soil and topographic maps, or field exploration. The size and spatial distribution of the land area does not affect the stratification of the project site. If two plots are far apart but meet our criteria for dividing the carbon layer, these two plots can still be seen as the same carbon layer.

The main purpose of carbon stratification is to combine plots of the same category, so as to obtain an accurate estimate of the carbon stock of the plot with a small sample size. Both the baseline investigation and the monitoring process may involve the division of carbon layer, and the basis for the two divisions is different, but the purpose remains the same.

Baseline carbon stratification is mainly based on land coverage and land use obtained from field exploration. Regarding land coverage, main considerations are on the type of land covering, the species, height, and coverage of herb and shrub; regarding the land use, mostly the current anthropogenic use on the land is considered, such as whether it is a barren hill and whether there is grazing. In addition, considering different altitude ranges and vegetation types, altitude also serves as a basis for the division of carbon layers.

According to the requirements of carbon stratification, detailed land coverage and land use survey form are used to record basic information such as the township, village committee, natural village, and ownership of the land; land coverage records the type of coverage, height, and canopy of vegetation, and information on the type, number, and age of scattered wood; land use records whether there are barren hills, the presence of grazing, grazing intensity, and grazing time. While identifying plot boundary, the questionnaire is filled out in detail according to field surveys and interviews, and each plot is photographed. All of these serve as basis for the division of carbon layer.

After carbon stratification, carbon layer and project plots should also be coded to fully comply with the standards set by the project. The coding rule is county name (alphabetical letter) + carbon layer code.

In this project, the baseline carbon layer is divided into three, as shown in Table 3.

(5) Project community survey

The community survey of the project is conducted in the same way as that in “the forestry carbon sink, community, and biodiversity project in Northwest Sichuan.”

(6) PDD development

The PDD development is conducted in the same way as that in “the forestry carbon sink, community, and biodiversity project in Northwest Sichuan.”

(7) Signing of land use contract

The land use contract for the project is signed in the same way as that in “the forestry carbon sink, community, and biodiversity project in Northwest Sichuan.”

(8) Project review

The project is a voluntary carbon trading project. It adopts over-the-counter trading model. Therefore, a support letter is not required from the NDRC, but it needs permission from forestry authorities and is kept on record in local forestry departments. In addition, in order to ensure that multiple benefits of the project are recognized, CCB review is required.

2. **Project construction period (2006–2009)**

During the project construction period, it is mainly necessary to carry out project construction in accordance with relevant national standards and PDD. Prior to project construction, the project afforestation design documents need to be approved by local forestry authorities and can only be implemented after the endorsement of relevant experts.

The main activities include soil preparation, afforestation, survival rate inspection and replanting, and forest land tending.

3. **Project Operation Period (2010–2037)**

After the project construction is completed, the project enters the daily operation and maintenance stage. This stage involves a series of activities to ensure the successful implementation of the project, such as daily care, fire and pest control, and certain thinning. In this project, two thinning activities were designed in the 10th and 15th years of the project, namely, in 2015 and 2022. The thinning intensity will be 30% of the total volume. The thinning shall be applied to the competent forestry authorities, and conducted in accordance with relevant national and local standards and subject to supervision by local forestry authorities.

In addition, during the operation period, sample plots will be laid out in accordance with the monitoring plan, and monitoring will be conducted once every 5 years. Monitoring is done by the owners themselves or inviting competent third parties. After the monitoring results are completed, DOE is invited to conduct an audit. According to the provisions of the CDM-EB, the DOEs that are generally required for verification and certification of the same project must be different, except for small-scale afforestation projects. Therefore, the project can select the same DOE for certification work. After the certification is completed, the DOE requests the CDM-EB to issue corresponding emission reduction.

In the project, the main project activities during the project period are described as follows: forest land tending, fire and pest control, thinning, monitoring, certification of emission reduction, and issuance of transactions.

4.2 *Difficulties and Solutions*

(1) Fire and pest risk

Fire and pests are potential risks in project activities. The project will reduce this risk through education and technical training for the community, strengthening patrolling and protection, and establishing fire prevention belts. In addition, mixed forest management models will also effectively reduce fire and pests. The project will invite local forestry bureaus to conduct training on fire awareness and fire responsibilities. At least eight forest guards will be hired from local community to conduct forest patrols. In local community, there are fire teams organized spontaneously by villagers. Once a fire breaks out, the team will participate in the fire extinguishing quickly and efficiently.

(2) Soil preparation

Soil preparation will cause disturbance to the vegetation and soil on the afforestation land. The main technical measures for alleviating the effect of soil preparation on vegetation and soil are to reduce the planting density (1667 plants/ha), prepare small planting holes (30–40 cm in diameter or 0.07–0.13 m^2 in area), and to maximize the existing vegetation. In this way, the ground surface area disturbed by soil preparation is estimated to only account for 1–2% of the total area. The planting holes are laid along the contour lines in a triangle shape to reduce soil erosion. Therefore, soil preparation and woodland clean-up will have less negative impact on soil and original vegetation.

(3) Fertilization

Fertilization in hole is used for the project instead of sowing, which can minimize the potential risk of fertilization.

(4) Pesticides

Unreasonable pesticide can be harmful to the natural environment, including contaminated soil, water, and air, and can be detrimental to the living environment of wild animals. In the project, environment-friendly measures will be adopted, including mixed tree species, quarantine of seeds and seedlings, and implementation of integrated pest and disease control measures. In particular, biological measures will be adopted to control pests and diseases, thereby limiting the use of pesticides.

(5) Changes in land use

Due to the fact that different landowners may have various expectations on land benefits, the project may be reversed in the future. For example, individual landowners hope to see large-scale benefits in the short term, so the project cycle is often too long for them; as project implementers, companies may fear that failure of project execution will result in the loss of their previous investment, so they hope land use can remain relatively constant during the project period; as the owners of state-owned land, because of the need to develop the economy, local governments do not want the project land to be occupied for too long. Therefore, when implementing the project, it is necessary to strengthen the education of landowners, highlight the ecological benefits of forestry construction, especially the outstanding contribution to guarding against natural disasters, and the long-term return of forestry. In addition, there needs to be contingency plan for possible reversal with the state-owned landowners during project implementation, such as finding alternative land.

5 Project Benefit

1. **Social benefits**

(1) Establishment of provincial and county forestry carbon management offices

In order to promote the project, with the support of the project, Forestry Department of Yunnan Province set up the Yunnan Provincial Forestry Carbon Management Office under the leadership of the head of the Forestry Department, and personnel were dispatched from the Yunnan Institute of Forestry Inventory and Planning to help with the management and coordination in the office. In addition, the Tengchong County Forestry Bureau also set up a forestry carbon management office, which consists of local forestry bureau and the Natural Reserves. The FCCB project purchased daily office equipment for the two carbon offices and conducted project management training.

(2) Establishment of strong implementation team

Before the project began, local partners learned about the concept of forest carbon trading for the first time. During the implementation of the project, personnel sent by forestry bureaus, natural reserves, enterprises, local forestry stations, and protection stations formed the implementation team. In addition, in order to increase the capacity of local implementation teams, the project invested heavily and provided manpower to train the implementation team in terms of climate change, carbon transactions, development and implementation of forestry carbon projects, project baseline investigation, project community surveys, project monitoring, GIS, remote sensing, and GPS applications.

(3) Enhancing social cohesion

Due to the long period of the forestry project and its slow effect, if there is no external support, it will be difficult for the community and farmers to implement the project and benefit from it. The project reflects the joint cooperation of governments, communities, natural reserves, forest farms, and NGOs. By integrating scattered farmers and communities to implement the project, these people have improved their capacity to respond to market changes and increased awareness of community cooperation.

(4) Technical training and demonstration

The results of community survey show that communities and farmer households often lack skills in obtaining high-quality source seeds, breeding high-yielding seedlings, and preventing fires and pests and diseases. This is also a major obstacle to farmers in local communities. In the project, local forestry systems and forest farms will organize training to help them understand the problems encountered in the implementation of project activities, such as seedling selection, nursery management, soil preparation, reforestation model, and integrated pest management.

2. **Economic benefits**

In order to maximize social and economic benefits, the project design process uses a participatory process. A participatory rural appraisal assesses the preferences, wishes, and concerns of local farmers by visiting and consulting farmers in the project area, so that the project can respond to their requests and help them improve their livelihoods. It is up to them to decide on the shareholding arrangements for farmers, communities, and forest farms that they are most willing to accept. It is estimated that 2108 villagers in 433 households of 5 villages, 3 townships will benefit from the project.

The main social and economic benefits of the project include the following.

(1) Increase income

According to the agreements reached between forest farms and local farmers, farmers will benefit from wood products and non-wood products obtained from concerned land. In addition, work compensation can also be obtained through the reforestation process.

(2) Sustainable fuelwood supply

Local residents, especially ethnic minority communities, have a certain degree of dependence on fuelwood. The project can provide local communities with more sustainable fuelwood resources.

3. **Environmental benefits**

(1) Biodiversity conservation

The project site is located in a global hotspot of biodiversity conservation and provides a suitable habitat for rare and endangered animals and plants. This shows the project has very outstanding benefits of biodiversity conservation:

A buffer zone was set up around the Gaoligong Mountains, and afforestation will gradually reduce the coverage of the invasive Eupatorium adenophorum, especially controlling its spread to natural reserves.

Provide community residents with necessary fuelwood resources, thereby saving the effort of community residents in logging from natural reserves and reducing the threat to biodiversity conservation.

The project will provide community residents with a continuously increasing source of income, which will effectively prevent community residents from illegal logging, poaching as well as the acquisition of unsustainable non-wood forest products in the natural reserves due to livelihood pressure.

In order to minimize the disturbance to vegetation, a belt-shaped soil preparation method and small planting holes were adopted. Natural regeneration will occur along with the growth of *Alnus cremas-togyne* and *Betula platyphylla*, so the project will contribute to close-to-nature forests and contribute to biodiversity conservation.

(2) Control of soil erosion

The parent rock formed in the project soil is granite or gneiss. Due to the high sediment content in the soil, the soil can be easily eroded. The afforestation will effectively protect the soil.

6 Experience and Lessons Learned

Land use and land use change contribute to 25% of global carbon emission. Therefore, forestry activities, such as afforestation/reforestation, will promote vegetation absorption of carbon dioxide, the main GHG in the atmosphere. In the past climate negotiations, the non-permanence of afforestation/reforestation may bring about potential risks of biodiversity loss and exacerbate community poverty, resulting in a large number of different opinions. How to balance opportunities and risks? We believe that a mechanism can be found to mitigate climate change through forestry activities, promote biodiversity conservation, and focus on sustainable community development. Therefore, we introduced CCB standards into project design and project implementation, which fully reflects the comprehensive benefits of forest ecosystems in these areas. In order to better promote the implementation of this standard and to set an example in the world, we have worked with the State Forestry and Grassland Administration and the Forestry Department of Sichuan Province to design an FCCB project since 2005. In the process of project implementation, some experience has been accumulated and some challenges have also been encountered.

1. **Experience**

Establish provincial- and county-level forestry carbon management offices to conduct overall management of the project. The project was strongly supported by the Forestry Department of Yunnan Province. The Bureau established a forestry carbon management office with the head of the Bureau as its director, and deployed staff to manage day-to-day business; office personnel reported to the afforestation division. The office and project team held regular meeting and set up a coordination mechanism, which is essential for solving problems in the project timely. In addition, Tengchong County also formed a forestry carbon management office consisted of county forestry bureau and natural reserves to support project implementation in Tengchong. In order to advance the project, the project team conducted lots of training for CDM A/R and CCB standards, and established a strong project implementing team at provincial and local levels. For most people, forest carbon trading and afforestation/reforestation activities under the CDM are completely new and novel. In the project design process, related activities need to be designed according to the CCB standards. For example, for baseline surveys, community surveys, and biodiversity surveys, it is necessary for them to comply with the requirements of CDM as well as with CCB standards. For this reason, the project team has devoted a great deal of time and energy to carrying out related training and elaborating on the rules and regulations. This is why the project garnered the world's first CCB gold medal.

In order to ensure the successful implementation of the project, the project team created a complete project management system. First, the annual meeting for key partners and stakeholders. Each year, the project team will hold regular annual meetings in Beijing to invite key partners and stakeholders involved in the project to participate. The meetings will sum up the work of previous year and plan and design for the next phase of the project. If the proposal of partner is adopted, it will be included in the annual work plan and receive financial support. Second, dedicated persons for tracking project work plan. One person will be assigned to manage the work plan that is agreed upon at the beginning of the year. The person will remind the implementing team of key nodes in the project, review the key outputs in the project, and have them re-examined by experts. Third, a complete project supporting mechanism. A complete supporting mechanism is in place regarding the work plan, meeting minutes, project documents, photos, videos, and PPT.

In the project implementation process, how to guarantee the enthusiasm of landowners in the project and balance the relationship between project implementing body and landowner is a crucial issue. If landowner does not voluntarily participate in the project, or has certain questions or even confrontation with the project, it will pose a serious threat to project completion. In addition, if the project implementing company does not have enough knowledge of the project, it will also encounter implementation problems. Therefore, in order to ensure the smooth implementation of the project, it is necessary to set up an exchange platform that all parties are satisfied with. All parties can fully express their opinions and appeals, and reach certain agreement. In the process of project design and project implementation, both parties should be fully involved. The rights and interests of landowners should be fixed through community meetings and communicated to each landowner. For project implementing companies, sufficient publicity and exchange must be ensured.

2. **Lessons**

During the implementation of the project, many challenges have been encountered. Especially in the initial stage of project implementation, due to the lack of knowledge of the rules, plus that the project concept was introduced for the first time, many participants misunderstand and misinterpret the project. Through project implementation, the lessons learned are as follows.

The stability of the project team must be maintained. A stable project team, especially a stable local implementation team, is essential. In the implementation of this project, due to the frequent replacement of personnel, the cost of personnel training has increased. The project implementing team is mainly responsible for communicating with the community. Because of personnel replacement, the established sense of trust needs to be re-established, and the lack of understanding by newcomers also contributes to the difficulty in communicating with the community.

The unclear landownership leads to more difficulty in implementing the project. The project was launched before the reform of collective forest ownership system. At that time, there was no clear demarcation on the collective forest boundary, causing increasing disputes over landownership among community residents. Such disputes increase the risk of scheduled project implementation. Therefore, the first priority in the project implementation is to clarify the land boundary with local forest owners.

The forestry project cycle is too long, and landowners and companies have insufficient understanding of the long-term effects of the project, resulting in low enthusiasm for investment in the construction period. For example, the project period is 30 years, while landowners' desire for land is benefits in the short term. However, the project is a unilateral type, which requires the project undertaking company to invest funding in afforestation through self-raising, borrowing, and government financing in the early stage. Since people are not optimistic about future carbon trading, if there is no carbon fund for afforestation in the earlier period, the enthusiasm of implementing company in implementing project will be greatly reduced. Since the forestry cycle is long, it must be ensured that the land use method remains unchanged during the project cycle after successful afforestation. The local government has some concerns about the project's long-term possession of land resources, especially if the land selected is located around developed transportation routes, the government may choose to change the land use due to the demand for economic growth. This also poses certain risks for future implementation of the project.

Public Voluntary Forest Carbon Project in China

Nuyun Li and Fangyi Yang

Summary

This chapter describes public voluntary forest carbon projects that aim to increase forest area and improve forest quality through afforestation and forest management. Such carbon projects have the objectives of increasing forest carbon, achieving ecological conservation, and raising community income. Such projects include enterprise, group or individual donation to public foundations, or entrusted social organizations and institutions for implementing afforestation projects that are in line with internationally accepted technical standards, and also include public awareness raising campaign, training courses, and research programs on climate change.

China's economy has grown rapidly and its GHG emission ranks the first in the world. Although China does not undertake the mandatory GHG emission reduction obligations under the Kyoto Protocol, as a responsible country of significance, the Chinese government has taken various measures to actively respond to climate change, such as improving energy efficiency, reducing energy consumption, using clean energy, and encouraging enterprises to actively participate in public welfare activities in response to climate change.

Since 2008, many NGOs have carried out various forms of voluntary forest Public voluntary forest carbon in China. We use the following examples, namely, the forest carbon projects by the China Green Carbon Foundation and the Beijing Shanshui Conservation Center, which are among the earliest Public voluntary forest carbon; and the massive ant forest project launched by the Ant Financial Group in recent years. These projects demonstrate the specific practices, successful experience, and results achieved by Chinese NGOs and companies with social responsibilities in conducting

N. Li (✉)
China Green Forest Carbon Foundation, Beijing, China
e-mail: linuyun516@vip.sina.com

F. Yang
Paradise Foundation, Beijing, China
e-mail: fangyi.y@139.com

Z. Lu et al. (eds.), *Forest Carbon Practices and Low Carbon Development in China*, https://doi.org/10.1007/978-981-13-7364-0_7

voluntary forest Public voluntary forest carbon. Based on the CDM A/R project methodology and other international rules on forest carbon of afforestation, as well as domestic reality, the China Green Carbon Foundation formulated a series of standards and rules, such as technical regulations for carbon afforestation, forest carbon project measurement and monitoring methodologies, and forest carbon project approval and verification regulations. It took the lead in launching projects such as carbon afforestation, carbon neutral projects, low-carbon tree-planting initiatives, and the dissemination of green carbon. All the afforestation projects and carbon neutral projects have been comprehensively managed from design to implementation, from carbon measurement to monitoring, from project review to registration, creating a standardized management of China's Public voluntary forest carbon for the first time. Through the pilot of voluntary carbon trading, it laid the foundation for promoting the marketization ecosystem of carbon.

In addition, since 2010, FAW-Volkswagen Audi Sales Dept. collaborated with the Beijing Shanshui Conservation Center to develop carbon projects in three nature areas in Liangshan Prefecture of Sichuan Province. They adopted "panda standard" to develop carbon projects and worked on FCCB recovery for giant panda habitat. The emission reduction of carbon dioxide generated will be used to offset the carbon emissions generated during the production and operation process of Audi, and to disseminate the concept of honest and transparent certified carbon emission reduction to the public. The project provided local communities with employment and labor skills training opportunities, while effectively improving the livelihood of local people. With the development of China's mobile Internet, it is possible to engage the public in low-carbon life and voluntary carbon reduction through mobile Internet tools. The Internet platforms are also becoming a trend to offset carbon emission for users through self-investment or the introduction of third-party funds. Among them, the "ant forest" initiated by the Ant Financial (an Internet company) and the "eco-coins" initiated by the Mobike are good examples.

1 Background of Public Voluntary Forest Carbon Projects in China

Public voluntary forest carbon China is a developing country with rapid economic growth. It ranks the first in the world in GHG emission and thus faces tremendous pressure from the international community to reduce GHG emission. As a responsible big country, according to the "common but differentiated" principle of the United Nations Framework Convention on Climate Change, China does not need to undertake mandatory GHG emission reduction obligations, and the country has not set quantity mitigation target for GHG emission of domestic enterprises, but the Chinese government has taken various measures to combat climate change, such as reducing energy consumption and using clean energy to reduce GHG emission, while encouraging enterprises to participate actively in charitable activities that contribute to forest emission reduction and combat climate change. The voluntary forest carbon project can not only help companies reduce their carbon footprint, but also assists companies and individuals in achieving broader social responsibility goals. After the international community reached the new agreement on global climate governance in 2015, namely, the Paris Agreement, the implementation of China's INDC target, in addition to the government's mandatory requirement that enterprises reduce GHG emission, the public and enterprises are required to actively participate in social service activities that contribute to carbon and reduce emission, so as to accelerate the intended INDC.

In addition, due to the design requirement of CDM project regime, it is required that forest carbons (carbon credits) to be purchased can only be resolved by issuing time-sensitive CERs to overcome non-permanence, resulting in low price of forest carbon in the CDM carbon market. Due to the complexity of its measurement and monitoring, forest carbons account for only a small portion of the CDM carbon market.

However, due to the relatively high positive social and environmental benefits brought about by forest carbon, many carbon buyers do not have the need for mandatory emission reduction in the voluntary market. The purchase of carbon credits stems more from corporate social responsibility and public welfare purposes. Forest carbon is more likely to be favored by buyers. When purchasing forest carbon, most buyers are willing to pay slightly higher price than the carbon credits from other emission reduction projects. Because forest carbon project can not only bring carbon emission reduction, the forest also has water conservation, water and soil conservation, and air purification, and multiple benefits such as ecological services and the promotion of local community development. At present, there are many enterprises and individuals who donate funds for forest carbon project for the purpose of corporate social responsibility and public welfare, and they participate in and experience the development of forest carbon project. These forest carbon projects do not contribute to afforestation through commercial purposes. Carbon measurements and monitoring were carried out on them and had relatively better social benefits. These can all be classified as voluntary forest carbon projects (Table 1).

As climate change has become the most important and pressing ecological and environmental issue both inside and outside the country, the development of Public voluntary forest carbon is speeding up. Compared with CDM carbon project, such project does not require complicated procedures such as method approval and verification in accordance with CDM rules, thus reducing costs. It also respects the intention of donors and allows them to participate in the process of planting forests. It meets the requirements of enterprises and the public, and is widely accepted. The rapid development of such project has also played a very good role in environmental education and climate change campaign.

Table 1 Voluntary forest carbon projects implemented by companies and social organizations during 2008–2017

Project name	Donating company and individual	Implementing location	Implementing agency	Afforestation area	Carbon sink methodology
Ant forest	Ant Financial Group	Alxa, Kubuqi, Inner Mongolia	SEE Foundation China Green Foundation etc.	N.A	
Inner Mongolia Shengle international ecological demonstration area, forest carbon project	Lao Niu Foundation	Horinger County, Inner Mongolia	China Green Carbon Foundation Lao Niu Foundation Inner Mongolia Autonomous Region Forestry Administration TNC	38,800 mu	CDM carbon afforestation methodology, CCB standard
China petrol carbon forest project	CNPC	21 provinces (district, municipality)	China Green Carbon Foundation	1.2 million mu	SFGA technical regulation on carbon afforestation, guidelines for the measurement and monitoring of carbon in afforestation projects
Lao Niu carbon forest carbon project for winter olympic games	Lao Niu Foundation and other agencies	Chongli, Chicheng, Huailai in Zhangjiakou, Hebei Province	China Green Carbon Foundation Lao Niu Foundation Hebei Provincial Department of Forestry Forestry Bureau of Zhangjiakou	31,000 mu	CCER carbon afforestation methodology
Kangbao ecological recovery engineering project	Spring Airlines, Spring International	Kangbao County, Hebei Province	China Green Carbon Foundation	4,011 mu	SFGA technical regulation on carbon afforestation, guidelines for the measurement and monitoring of carbon in afforestation projects

(continued)

Table 1 (continued)

Project name	Donating company and individual	Implementing location	Implementing agency	Afforestation area	Carbon sink methodology
Afforestation Carbon projects in Dongjiangyuan, Hong Kong Jockey Club (4 phases in total)	Hong Kong Jockey Club	Longchuan County, Guangdong Province	China Green Carbon Foundation	8,000 mu	SFGA technical regulation on afforestation carbon projects, Guidelines for the Measurement and monitoring of carbon in afforestation projects
Changlong carbon afforestation project, Guangdong Province	Chimelong in Guangdong	Heyuan and Meizhou, Guangdong Province	China Green Carbon Foundation	1.3 mu	CCER carbon afforestation methodology
Carbon neutral project in COP16, Tianjin in 2010	China Guodian, Lu'an Environ-mental Protection and Energy Develop-ment Company of Shanxi	Xiangyuan, Xiyang, Pingshun counties, Shanxi Province	China Green Carbon Foundation Forestry Department of Shanxi Province	5,000 mu	SFGA technical regulation on afforestation carbon, Guidelines for the Measurement and Monitoring of Carbon in Afforestation Projects
2014 APEC Summit meeting, carbon neutral project	CITIC Group, Spring Airlines	Huairou District, Beijing; Kangbao County, Hebei Province	China Green Carbon Foundation Beijing Gardening and Greening Bureau	1,274 mu	SFGA technical regulation on afforestation carbon, guidelines for the measurement and monitoring of carbon in afforestation projects
2016 G20 Hangzhou Summit, carbon neutral project	Lao Niu Founda-tion, Wanma Group	Lin'an, Zhejiang Province	China Green Carbon Foundation Forestry Department of Zhejiang Province Hangzhou Municipal Government	334 mu	SFGA technical regulation on afforestation carbon, guidelines for the measurement and monitoring of carbon in afforestation projects

(continued)

Table 1 (continued)

Project name	Donating company and individual	Implementing location	Implementing agency	Afforestation area	Carbon sink methodology
Carbon neutral project of Annual Summit of China Green Companies (7 sessions)	Lao Niu Foundation	Horinger County, Inner Mongolia	China Green Carbon Foundation	404 mu	SFGA technical regulation on afforestation, carbon guidelines for the measurement and monitoring of carbon in afforestation projects
Panda forest project	FAW Audi	Liangshan, Sichuan Province	Beijing Shanshui Conservation Center	2,000 mu	Refer to CCB standard
	Uni-President Enterprises Corporation	Liangshan, Sichuan Province	Beijing Shanshui Conservation Center	1,000 mu	Refer to CCB standard

2 Exploration and Practice of China Green Carbon Foundation

The China Green Carbon Foundation is the first national public foundation for the purpose of increasing forest carbon, reducing emission, and combating climate change. It is also currently the most authoritative and professional organization for public activities to implement carbon compensation, carbon neutrality, and other measures through afforestation, forests protection, and carbon emission reduction.

2.1 The Establishment of the China Green Carbon Foundation

On July 19, 2010, with the approval of the State and the registration with the Ministry of Civil Affairs, China Green Carbon Foundation (hereinafter "the Carbon Foundation"), which is China's first national public foundation, was established with the aims to increase forest carbon, reduce carbon emission, and tackle climate change. Its predecessor was the China Green Carbon Fund established in 2007. In 2007, China National Petroleum Corporation (hereinafter CNPC), in cooperation with the State Forestry and Grassland Administration, launched an afforestation carbon project with the main objective of increasing forest carbon and reducing carbon emission.

It donated RMB 300 million on July 20 of the same year to set up the China Green Carbon Fund (as a special fund at that time and was temporarily put under the China Greening Fund). On August 3, 2010, the inaugural meeting of the China Green Carbon Foundation was held in the Great Hall of the People in Beijing, which marked the transformation of the China Green Carbon Fund from a special fund into a national public foundation. The Foundation was initiated by the CNPC and its work is subject to the State Forestry and Grassland Administration.

2.2 *Main Working Scope of China Green Carbon Foundation*

According to the "Charter of China Green Carbon Foundation", the main working scope of the Foundation is: to carry out afforestation, forest management, desertification control, fuel forest base construction, wetlands, and biodiversity conservation for the purpose of addressing climate change; to create a variety of memorable forests for the purpose of accumulating forest carbon, and to develop activities such as adoption of green land; to strengthen the protection of forests and woodlands and to reduce carbon emission from irrational land use; to support all kinds of scientific and technological research and educational training for the purpose of increasing forest carbon and reduction emission; to develop carbon measurement and monitoring and to develop relevant standards; to actively promote the function and role of forests in combating climate change; to raise public awareness of protecting the ecological environment and paying attention to climate change; to conduct domestic and international cooperation and exchange on combating climate change; and to carry out other public activities suitable for the purpose of the Foundation (Fig. 1).

2.3 *Operation Mode of the Carbon Foundation*

China's forestry takes innovative attempts to cope with climate change, such as in the form of public fundraising, mobilizing social forces and carrying out carbon increase and emission reduction activities aiming at addressing climate change. According to the central task of China's national strategy for addressing climate change and the development of modern forestry, the Carbon Foundation combines the progress and development trend of forestry issues in the international negotiation process on climate change, shoulders the mandate of "increasing vegetation, absorbing carbon dioxide, addressing climate change, and protecting the earth homeland," and operating with the mode of "standard operation, scientific measurement, and openness under the principle of public welfare." It has carried out beneficial exploration and practices in project implementation and standardized management, and has achieved remarkable results. The Foundation's operating framework is shown in Fig. 2.

In accordance with the *Regulations on the Management of Foundations* by the State Council and relevant laws and regulations, the Carbon Foundation formulated

Fig. 1 Inaugural meeting of the China Green Carbon Foundation

the *Administrative Measures for the Funding Management of China Green Carbon Foundations*, the *Administrative Measures for Project Management of China Green Carbon Foundation*, *Administrative Measures for Financial Management of China Green Carbon Foundation,* and *Administrative Measures for Special Fund Management of China Green Carbon Foundation*, as well as other rules and regulations. Through forestry management, it is committed to building a public platform for enterprises and the public for the purpose of "storing carbon credits, fulfilling social responsibilities, increasing farmers' income, and conserve the ecology." The China Green Carbon Foundation strictly supervises its management. Companies donated funds to the China Green Carbon Foundation. All afforestation needs carbon measurement and registration. All projects are subject to contract management and monitored throughout the entire process. They are monitored by all parties and audited by relevant departments to ensure that funds are in place, operations are smooth, and ecology and social benefits are significant. The forests that have been planted are owned by farmers or are for public welfare. They play an important role for the industry to feedback agriculture and the city to feedback the countryside. Farmers got employment opportunities and increased their income by participating in afforestation, while donating enterprises obtained carbon credits (credit indicators) that are measured according to regulations, which is recorded in corporate social responsibility accounts. The implementation of the project increased forest vegetation and forest quality; meanwhile, it also contributes to social, ecological, and economic benefits in absorbing and storing carbon dioxide, releasing oxygen, protecting biodiversity, conserving water resources, preventing wind and sandstorm, providing by-products

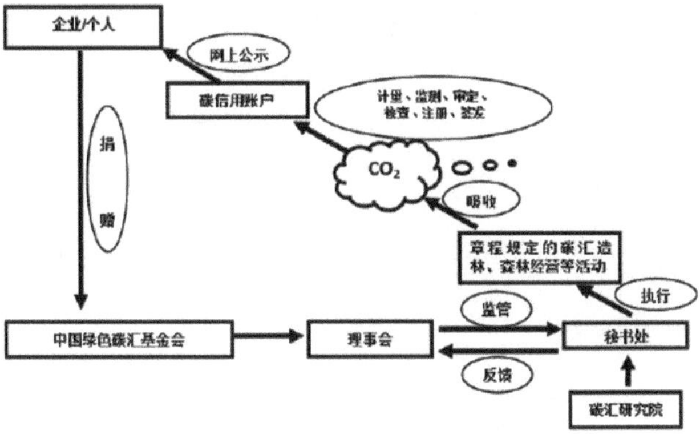

企业/个人	Enterprise/individual
捐赠	donation
中国绿色碳汇基金会	China Green Carbon Foundation
网上公示	Online notification
碳信用账户	Carbon credit account
理事会	Board
计量、监测、审定、核查、注册、签发	Measurement, monitoring, validation, verification, registration, certification
吸收	Absorption
章程规定的碳汇造林、森林经营等活动	Forest Carbon , afforestation, forest management as specified by the Charter
监管	Supervision
反馈	Feedback
执行	Implementation
秘书处	Secretariat
碳汇研究院	Forest Carbon research institute

Fig. 2 Operating framework of China Green Carbon Foundation

of forests, and good recreational sites. It made positive contributions to safeguarding national ecological security, promoting green growth, and combating climate change. In addition, many individuals also actively participated in the afforestation to mitigate global warming. They donated money to the China Green Carbon Foundation to "purchase carbon credit," in order to offset the carbon dioxide emitted in their daily lives. "Participate in carbon offsets and eliminate carbon footprints." Similarly, all individuals who donate money for afforestation also perform carbon measurement and monitoring, and register with their respective "carbon credit" account, which is publicized on the China Green Carbon Foundation's website (Fig. 3).

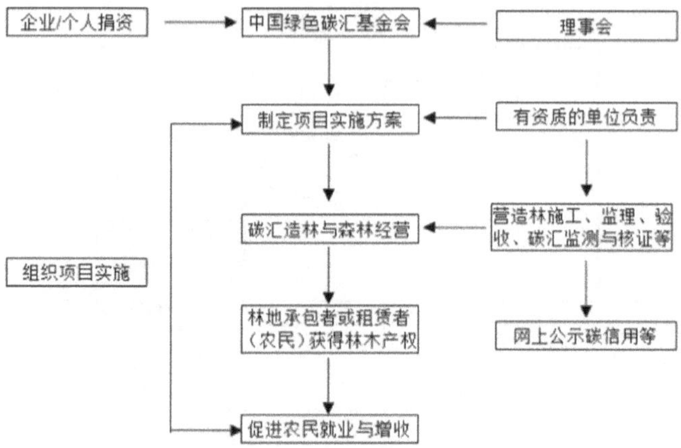

企业/个人捐资	Corporate/individual donation
组织项目实施	Project implementation
中国绿色碳汇基金会	China Green Carbon Foundation
制定项目实施方案	Draft project implementing plan
碳汇造林与森林经营	afforestation and forest management
林地承包者或租赁者（农民）获得林木产权	Contractor or leaser obtain property rights of forest product
促进农民就业与增收	Promote farmerĺs employment and raise income
理事会	Board
有资质的单位负责	Undertaken by certified units
营造林施工、监理、验收、碳汇监测与核证等	Implementation, supervision, commission, carbon monitoring and validation etc.
网上公示碳信用等	Online of carbon credit

Fig. 3 Operating chart of China Green Carbon Foundation

2.4 Public Voluntary Forest Carbon Projects by the Carbon Foundation

According to the willingness of donor and the business scope of China Green carbon foundations, the Foundation mainly supports the following projects with its funding, namely, afforestation carbon, publicity campaign, education, and training aimed at promoting public awareness.

(I) Carbon project with afforestation
1. Definition of carbon project with afforestation

According to the *Technical Regulations on Carbon and Afforestation (Trial version)* by the State Forestry and Grassland Administration (SFGA 2010) and *Guidelines for the Measurement and Monitoring of Carbon in Afforestation Projects* (SFGA

2011a, b), afforestation is an activity of special requirement that is done on soil with identified baseline, with the main purpose of increasing carbon, and conducting carbon measurement and monitoring with respect to afforestation and forest (wood) growth processes.

The main differences between afforestation projects for carbon and conventional afforestation types are as follows:

(1) The baseline scenario needs to be determined. The baseline scenario refers to the land use method that can reasonably represent where there is no proposed afforestation project, given the technical conditions, financing capacity, resource conditions, policies, and regulations of the project area.
(2) The plots for afforestation project for carbon plots shall meet the requirements for land eligibility under the *Technical Regulations on Afforestation project for carbon (Trial version).*
(3) The net carbon generated by afforestation projects shall be measured, monitored, reviewed, verified, registered, and certified for issuance.
2. Plot selection of afforestation project for carbon.

When the Carbon Foundation implements afforestation projects for carbon, based on the willingness of donors and the actual distribution of suitable forest lands, and in consultation with relevant provincial (district, city, county) forestry departments and the requirement of *Technical Regulations on Afforestation project for Carbon (Trial version)* shall give priority to areas where ecological location is important and where ecology is fragile and relatively poor. The barren mountain, wasteland, and marginal land with advantageous forest are selected as project area. At the same time, it is necessary to consider local biodiversity conservation, prevention of soil erosion, and promotion of local economic and social development. For example, in the forestation carbon project at Fangshan District in Beijing selected land that is ecologically fragile and mountainous area that is in urgent need of greening (Fig. 4).

3. Operational design and carbon measurement of afforestation projects for carbon afforestation projects for carbon
(1) Operational design of afforestation project for carbon

After the afforestation plots and afforestation tasks are identified, a forestry planning and design units with relevant qualification will carry out afforestation design in accordance with the *Technical Regulations on Afforestation project for Carbon (Trial version)* issued by the SFGA. Afforestation species, planting density, soil preparation requirements, planting time, planting technology, tending, management and protection, and other related technical measures were implemented on respective mountains, plots, and subplots.

(2) Carbon measurement and monitoring plan for afforestation projects

While contracting a forestry planning and design unit to carry out operational design of carbon project, a qualified forest carbon measurement and monitoring units, which

Fig. 4 Land use before implementing afforestation project for carbon in Fangshan district of Beijing

is contracted by the Carbon Foundation and approved by the SFGA, will carry out carbon measurement (prediction) and draft a monitoring plan, thus resulting in a Project Design Document (PDD).

According to the *Provisional Method for the Measurement and Monitoring of Forestry Carbon and Management* by the SFGA (SFGA 2011a, b), 10 institutes have obtained the qualifications for forestry carbon measurement and monitoring from SFGA, namely, Chinese Academy of Forestry, the Academy of Forestry Investigation and Planning of SFGA, the Planning and Design Institute of Forest Products Industry of SFGA, Kunming Forestry Prospecting and Design Institute of SFGA, Zhejiang A&F University, Inner Mongolia Agricultural University, Beijing Forestry Society, Beijing Forestry University, Nanjing Forestry University, and China Academy of Agricultural Sciences. In 2013, the number of qualified units increased to 15.

The measurement and monitoring unit intervened during operational design stage, so as to help the designing unit understand the requirements of carbon sink measurement and monitoring, verify land conformity, and follow the *Guidelines for the Measurement and Monitoring of Carbon Sink in Afforestation Projects* issued by the SFGA to conduct carbon sink measurement for afforestation projects for carbon, draft carbon measurement reports and monitoring plan, and later finalize the PDDs.

In order to ensure the measurable, reportable, and verifiable nature of the project's net carbon generated from afforestation projects, a monitoring plan must be developed while conducting carbon measurement. The measurement and monitoring unit shall strengthen contact and cooperation with the specific implementing unit or project owner to prepare a scientific and feasible monitoring plan. The monitoring plan includes monitoring content, monitoring methods, implementation plan, and precision control.

4. Validation of afforestation project for carbon

In order to obtain a true and reliable net carbon for the afforestation project for carbon and ensure its smooth entry into carbon trading system as carbon credit, the Research Institute of Forestry Policy and Information, Chinese Academy of Forestry, which is contracted by the Department of Afforestation and Greening, SFGA and funded by the Carbon Foundation, The *Guideline for the Validation and Verification of China's Forestry Carbon Project (Trial version)* (hereinafter the "*Guideline for the Validation and Verification*") was approved by an experts committee of the Carbon Foundation in April 2011 (China Green Carbon Foundation 2011a, b).

According to the *Guideline for the Validation and Verification*, after conducting the operational design and carbon measurement report, and drafting the carbon monitoring plan the carbon monitoring plan has been drafted, and the Carbon Foundation commissions a Validation and Verification Agency for Forestry Carbon Project (a third-party agency) to conduct the validation work. The validation body will evaluate project design, project land eligibility, scientific rationale of carbon measurement method, the rationale of carbon measurement results, and the feasibility of monitoring plan. Based on field inspections and textual assessments, a project will be identified as qualified carbon project and be issued a project validation statement. The statement is a critical basis and key link for the project to register and carry out carbon credit verification. The Carbon Foundation commissioned the Forest Green Asset Management Center under the Chinese Academy of Forestry to review the projects that have already been implemented and issue validation statement.

5. Registration of afforestation project for carbon

The registration of a forest carbon project is a process in which a registrar identifies a qualified carbon project that has been validated by a third-party agency as a certified carbon project and keeps it on the record. This is a key process to ensure the authenticity, reliability, and additionality of carbon credits generated by the forestry carbon project. It will also contribute to ensuring the equitability, openness, and fairness of forestry carbon transactions, safeguarding the legitimate rights and interests of all stakeholders in afforestation project for carbon and promoting sound and orderly development of China's forest carbon market. In accordance with the National Strategy for Addressing Climate Change and the *Action Plan for Combating Climate Change in Forestry*, the Carbon Foundation drafted in 2011 the *Provisional Management Methods for the Registration of Forestry Carbon Project of the China Green Carbon Foundation* (China Green Carbon Foundation 2011a, b), and established a registration platform for forest carbon project. The platform is located in the planning institutes of the SFGA. It is jointly managed by the Carbon Foundation and the planning institutes to carry out registration services for forest carbon project.

6. Implementation of afforestation project for carbon

There is no fundamental difference between afforestation projects for carbon and other afforestation projects. The soil preparation and plantation are based on the specific requirements of operational design. However, since the afforestation project for carbon by the Carbon Foundation is aimed at increasing carbon it is needed to measure and monitor the emission from all production activities and increased carbon. To this end, the *Technical Regulations on and Afforestation project for carbon (Trial version)* makesthe following requirements: soil preparation on cultivated land and controlled mountain burning are prohibited; primary scattered forests at the afforestation sites shall be protected, and shrubs or herbs shall be preserved as much as possible; no soil preparation shall be allowed in the nature reserves for minimum species, rare and endangered plant and animal, and buffering protection zones shall be kept with appropriate width; promote the use of organic fertilizers; tending shall be in a timely manner; forest fire prevention and pest control measures shall be implemented to maintain the soundness; stability of forest stands to reduce carbon emission; and pest and disease occurred during or after afforestation shall adopt comprehensive prevention and control measures based on biological control.

From 2007 to the end of 2016, the Carbon Foundation received nearly RMB 800 million in donation from enterprises and the public. It has created and participated in the management of over 1.2 million acres of forestland for carbon in 21 provinces (autonomous regions and municipalities).

7. Inspection and acceptance of afforestation project for carbon

During the afforestation period, the project management unit shall inspect and supervise all operational procedures at any time, and strictly follow the technical standard specified for the operational design to reduce carbon emission. The afforestation survival rate shall be checked 1 year after the completion of afforestation or one growth season, the forest acceptance and afforestation retention rate shall be checked after 3 years of afforestation. Among these, the operational design shall be verified on each subplots, and the area of afforestation and the survival rate of afforestation will be checked to see whether they are implemented according to operational design. The condition of unviable forestry pests and the implementation of mixed forests shall be checked. Based on the qualified afforestation area, the acceptance rate of afforestation, comprehensive afforestation indicators, comprehensive afforestation acceptance rate and afforestation retention rate (Figs. 5, and 6) afforestation project for carbon will be evaluated, and inspection and acceptance reports will be issued. Projects funded by the Carbon Foundation are all subject to inspection and acceptance.

8. Management of afforestation project for carbon

In order to ensure the safety of afforestation project for carbon during the crediting period, achieve multiple benefits such as increased carbon, improved ecological environment, and increased farmer income, the project implementing units have signed contracts with individuals or rural collective organizations in the form of management contracts. A designated person will be identified to ensure the implementation of daily forest fire prevention, pest control, and prevention of human and animal destruction.

Fig. 5 Survival rate inspection after 1 year of the afforestation project for carbon in Qingyang, Gansu Province

9. Monitoring of afforestation project for carbon

The Carbon Foundation signs contracts with agencies that have the qualifications for forestry carbon measurement and monitoring to undertake the monitoring of afforestation projects. According to the *Guidelines for the Measurement and Monitoring of Carbon in Afforestation Projects* and the project carbon monitoring plan and monitoring method, the carbon measurement and monitoring agencies collect data on project activities, changes in project carbon stock, greenhouse gas emission within project boundary, and leakage monitoring during the crediting period of the project. The relevant data required for monitoring shall be set up and a fixed sample plot shall be established. Sample plots are usually set up during the 5th, 10th, 15th, and 20th year of project implementation, so as to calculate the actual net carbon sink generated by the project and submit periodic monitoring report.

10. Verification of afforestation project for carbon

In order to ensure the accuracy and credibility of the monitoring results and ensure that forest carbon project smoothly merge into forestry carbon trading system, the *Guideline for the Validation and Verification* stipulates that carbon project shall, in addition to drafting carbon measurement report before implementation, conduct monitoring after project implementation, usually one monitoring report every 5 years.

Fig. 6 Acceptance staff conducting onsite inspection at the afforestation project in Qingyang, Gansu Province

Meanwhile, it is subject to the review of an independent third party. The amount of carbon verified by an independent third party is the basis for the issuance of carbon credits by forestry carbon project registration agency.

After project owner or implementing agency submits phased monitoring report, forest carbon verification is undertaken by a third-party verification agency to independently assess whether the monitoring method is scientific and reasonable and to evaluate the accuracy, reliability, transparency, reservation, uncertainty, and quality assurance. Based on the assessment results, the net carbon sink generated by carbon project is verified and confirmed, thus providing a basis for carbon project registration agency to issue carbon credits.

At present, the verification of afforestation project for carbon is contracted to the Forest Green Carbon Asset Management Center under the Research Institute of Forestry Policy and Information, Chinese Academy of Forestry, Beijing Zhonglin Green Carbon Sink Asset Management Co., Ltd., and other units. The development process of afforestation project for carbon is shown in Fig. 7.

Through the above scientific and rigorous processes, China Green Carbon Foundation achieved all-round management from project design to implementation and from project review to registration. This ensures the quality of the project and the reliability of the carbon generated. It ensures social, ecological, and economic benefits of afforestation project for carbon and lays the foundation for such projects to be integrated into national voluntary carbon emission trading system.

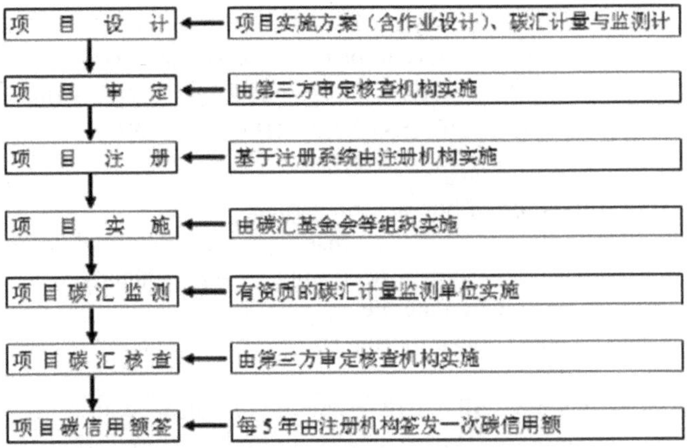

项目设计	Project design
项目审定	Project validation
项目注册	Project registration
项目实施	Project implementation
项目碳汇监测	Project carbon monitoring
项目碳汇核查	Project carbon verification
项目碳信用额签	Project carbon credit issuance
项目实施方案（含作业设计、碳汇计量与监测计量）	Project implementation plan (including operational design, carbon measurement and monitoring measurement)
由第三方审定核查机构实施	Implemented by a third-party validation and verification agency
基于注册系统由注册机构实施	Implemented by registration agency based on registration system
由碳汇基金会等组织实施	Organized and implemented by the Carbon Foundation
有资质的碳汇计量监测单位实施	Implemented by a qualified carbon measurement and monitoring unit
由第三方审定核查机构实施	Implemented by a third-party validation and verification agency
每 5 年由注册机构签发一次碳信用额	Carbon credit issued by a registration agency every five years

Fig. 7 Genera procedure of carbon sink afforestation project

(II) Carbon neutral project

Another type of carbon project undertaken by the Carbon Foundation is carbon neutral project.

1. Definition of carbon neutrality

The so-called carbon neutrality refers to the calculation of the total amount of GHG emission directly or indirectly generated by a company, organization, or individual within a certain period of time. It then passes the "carbon credits purchase" (i.e., the emitters invest in afforestation to increase or reduce carbon emissions) to offset its GHG emission, thus reducing the concentration of GHGs in the atmosphere.

The carbon emission of a company, organization, or individual is also called carbon footprint. It mainly refers to the total carbon emission generated by energy or resource consumption in production, operation, and living. It generally includes electricity, gas, transportation, etc. The unit is usually expressed in tons of carbon dioxide equivalent.

2. General procedures for carbon neutral project

(1) Measure carbon footprint (carbon emission). A professional carbon inventory agency will carry out activities according to the company's production and organization, and then calculate the amount of greenhouse gas emission based on energy and resource consumption, and issue a carbon footprint measurement report.
(2) A professional agency will calculate the area of afforestation needed to offset these carbon emissions. The agency will choose afforestation sites based on the willingness of donors and actual conditions, design afforestation models, estimate the afforestation areas required for carbon neutral emission, and issue carbon neutral reports.
(3) The Carbon Foundation organizes carbon neutral forest. The procedures for implementing the project are generally the same as those for afforestation project for carbon.

3. Introduction to major carbon neutral projects.

Since 2010, the Carbon Foundation has used the method of afforestation to increase forest carbon and organized 47 carbon neutral projects including large-scale international and domestic conferences, international organizations, companies, or related activities. These efforts effectively promoted the concept of green and low carbon, and made positive contribution to raising the awareness and ability of the public to tackle with climate change.

(1) On October 4–9, 2010, the 12th meeting of the ad hoc working group on long-term cooperation initiative of the UNFCCC and 14th meeting of the ad hoc working group on further commitment under the Annex A Parties to the Kyoto Protocol (hereinafter the COP16) were held at the "Meijiang Convention and Exhibition Center" in Tianjin, China. 4,000-odd representatives from more than 190 parties of the UNFCCC attended the meeting. This is the first time that the Chinese government hosted a meeting of the UNFCCC. The relevant national authorities decided to make this meeting a "carbon neutral" international conference.

As measured by the Institute of Energy Economics and Environment of Tsinghua University, the meeting's carbon emission was approximately 12,000 tons of carbon dioxide equivalent. The Planning and Design Institute of Forest Products Industry of the SFGA, as contracted by the Carbon Foundation, calculated that 5,000 mus of forest would need to be planted so that the above emission from the meeting could be fully absorbed within the next 10 years. The Carbon Foundation invested

Fig. 8 "Carbon neutralization" ceremony for the COP16 in Tianjin

RMB 3.75 million to organize the plantation of 5,000 mus of forests in the counties of Xiangyuan, Xiyang, and Pingshun, Shanxi Province. Farmers in the project are expected to receive RMB 2.6 million in labor income and more than RMB 7 million in forest by-products and timber revenue. The afforestation project was completed in early December of that year and a "carbon neutralization" ceremony was held (Fig. 8).

(2) APEC carbon neutralization project (2014)

In the morning of November 3, 2014, the launching ceremony of carbon neutralization plantation for the Asia-Pacific Economic Cooperation (APEC) meeting was held in Yanqi Town, Huairou District, Beijing. The ceremony was sponsored by the Ministry of Foreign Affairs, the State Forestry and Grassland Administration, and the Beijing People's Municipal Government. More than 100 representatives from related authorities, FAO, UNEP, and donor companies attended the launching ceremony and tree-planting activities. As measured by Beijing Gloriam Climate Technology Consulting Co., Ltd. and reviewed by the Carbon Foundation, the total carbon emission amounted to 6,371 tons of carbon dioxide equivalent. The Carbon Foundation and Beijing forestry departments organized China CITIC Group Co., Ltd. and Spring Airlines Co., Ltd. to donate RMB 6.5 million to plant 1,274 mus of forest in Huairou District of Beijing and Kangbao County of Hebei Province, among which Beijing accounted for 674 mus, and Hebei 600 mus, the main afforestation tree species are *Pinus tabulaeformis*, *Pinus bungeana*, *Pinus sylvestris, Platycladus orientalis*, *Astragalus, Pentagonal Maple*, *Eucalyptus,* etc. In the next 20 years, all the carbon emission generated by the meeting can be absorbed.

(3) Carbon neutralization project for the Annual Summit of China Green Companies (2011–2017)

From April 21 to 22, 2011, the Annual Summit of China Green Companies of the China Entrepreneur Club was held in Qingdao. The organizer decided to use the method of afforestation to absorb carbon dioxide and make the summit a green, low-carbon, environment friendly "carbon neutral" meeting. As measured by Beijing Gloriam Climate Technology Consulting Co., Ltd., nearly 800 delegates attending the summit generated 65.5 tons of carbon dioxide equivalent during transportation, accommodation, catering, and equipment use. With the donation from the Lao Niu Foundation, the China Green Carbon Foundation planted a forest of 53 mus in the ecologically fragile area of Horinger County of Inner Mongolia (the species planted are *Pinus sylvestris. mongolica*). The carbon emission from this summit can be absorbed in the next 5 years, thus achieving the "carbon neutral" of the summit. The successful implementation of this forest made the Annual Summit of China Green Companies in 2011 the first "carbon neutral" meeting of its kind since the inception. The subsequent 2012–2017 summits were all undertaken by the China Green Carbon Foundation to implement the carbon neutralization projects by planting trees in Horinger, Inner Mongolia. As of 2017, the Lao Niu Foundation donated a total of RMB 1.459 million to the summit for carbon neutral forest projects. By the end of 2017, the China Green Carbon Foundation and the Forestry Bureau of Horinger County, Inner Mongolia have established seven plots of the "Carbon Neutral Forest for the Annual Summit of China Green Companies" in the county, and planted suitable species *(Pinus sylvestris*) totaling 33,500 and covering 404 mus. The survival rate and conservation rate of afforestation are both over 87%, and the trees are growing in good condition. In addition to achieving the goal of "carbon neutrality" at each annual summit, the project is also conducive to increasing the protection of forest vegetation in ecologically fragile areas, improving ecological environment, and mitigating and adapting to climate change. At the same time, it also reduces the sandstorm in Beijing and Tianjin areas, promotes biodiversity conservation, increases the income of farmers and herdsmen, and safeguards ecological security. This public project has played an important role in guiding green and low-carbon development, advocating all sectors of society to practice low-carbon conference, low-carbon production, low-carbon administration, low-carbon life, and actively participating in building ecological conservation and beautiful China.

(4) NBAR carbon neutralization project (2010)

The International Network for Bamboo and Rattan (INBAR) has established a carbon offset fund within the organization. In accordance with the measurement standard of the China Green Carbon Foundation, a proportionate percentage of the cost will be drawn from the employees' travels and donated to the China Green Carbon Foundation. The carbon dioxide absorbed through afforestation will be used to offset the organization's official travel for the year. It is estimated that INBAR generated 61 tons of carbon dioxide equivalent in greenhouse gas in 2010. The China Green Carbon Foundation has used the donation of RMB 11,000 of INBAR to organize a project of 7.5 mus of new bamboo forest in Lin'an, Zhejiang Province to offset the 61 tons of carbon dioxide equivalent in greenhouse gases generated by the organization in 2010. The beneficiaries of this project are farmers in the local community. All afforestation projects were completed by the end of October 2012. The purpose is to offset the carbon emission caused by INBAR (such as business trips) in 2010 through the implementation of forest carbon project to realize the carbon neutralization goal of the organization in terms of business trip.

(5) "Green Singing · Zero Carbon Music Season" carbon neutralization project (2011)

From July to December 2011, the carbon emission from more than 130 performances organized at the Forbidden City Concert Hall, Beijing was fully absorbed by the China Green Carbon Foundation with afforestation investment. This is the second zero-carbon music season since the New Year Day from last year. It is estimated by a professional agency that about 900 tons of carbon dioxide equivalent will be generated during the entire music season. The goal of "zero emission" during the music season was achieved by developing the Beijing demonstration for forest to offset the carbon emissions generated in Badaling and Shenjiaying of Yanqing. Promoting ecological civilization, advocating public elimination of carbon footprint, participating in carbon offset, and striving to build a Beijing with humanity, science, and technology and a green Beijing to combat climate change and develop low-carbon cities.

From 2010 to 2017, the China Green Carbon Foundation organized and implemented 45 carbon neutral projects for large- and medium-sized meetings (or events) in China or abroad (see the Table 2), creating yet another "China First" (available for checking at www.thjj.org).

(III) Low-carbon plantation day project

Since its implementation in 1981, China's universal voluntary tree-planting campaign has become a public event with the widest participation, longest duration, and most significant and far-reaching impact in the world. For more than 30 years, the Central Party Committee, the State Council, and local leaderships at all levels have taken the initiative to engaging in the obligation of tree planting. The people of all ethnic groups across the country have responded actively and participated extensively. The voluntary tree-planting activities have made great achievements. By the end of 2015, a total of 15 billion people participated in the voluntary tree planting of about 70 billion trees nationwide, which amount to afforestation area of about 46 million hectares. The accomplished green coverage rate and green space rate across the country reached 40% and 36%, respectively. The nationwide voluntary tree-planting campaign has accelerated the greening of land, improved the ecological environment, promoted ecological civilization, and increased forest carbon stock, thus playing an active role in combating climate change. However, since the Reform and Opening up more than 30 years ago, rural areas have implemented "land contracting" system. Especially since 2008, China has carried out reforms of the collective forest ownership. Most of the land and collective forest land have been allocated to individual farmers for long-term use, but the barren mountains and wasteland that are suitable for forest vegetation recovery are mostly located in remote mountainous areas, with poor conditions, shortage of water resources, and few suitable tree species, thus it is difficult for afforestation and requires specialized expertise. Rural people who are nonprofessionals can go to plant trees, but this will

Table 2 List of carbon neutral projects of the China Green Carbon Foundation

No.	Project
1	Carbon neutral project for COP16 in Tianjin in 2010
2	Carbon neutral project for the 3rd Conference of China Ecological Civilization and Green Competitive in 2010
3	Carbon neutral project for the INBAR in 2010
4	Carbon neutral project for the "green singing and zero emission music season" in 2010
5	Carbon neutral project for the meeting of the head of national forestry agencies in 2011
6	"2010 carbon neutral enterprise" project of Jianfeng Packaging Co., Ltd. of Luojiang in Quanzhou, Fujian
7	Carbon neutral project for the Annual Summit of China Green Companies in 2011
8	Carbon neutral project for the working conference on fire prevention and control during autumn and winter of 2011
9	Carbon neutral project of public welfare for the INBAR in 2011
10	Carbon neutral project for the "green singing and zero emission music season" in 2011
11	Carbon neutral project for the meeting of the head of national forestry agencies in 2012
12	Carbon neutral project for the Annual Summit of China Green Companies in 2012
13	Carbon neutral project for official travel of the China Green Carbon Foundation
14	Carbon neutral project for the conference on low-carbon tourism in 2012
15	Carbon neutral project for the Annual Summit of China Green Companies in 2013
16	Carbon neutral project for Taihu World Cultural Forum in 2013
17	Carbon neutral project for the UN forum on sustainable consumption
18	Carbon neutral project for The First Shenzhen International Low Carbon City Forum in 2013
19	Carbon neutral project for the 5th conference on ecological conservation and green competitiveness in 2013
20	Carbon neutral project for the 1st training course on forestry management in 2013
21	Carbon neutral project for the Annual Summit of China Green Companies in 2014
22	Carbon neutral project for the Eco Forum Global Annual Conference Guiyang 2014
23	Carbon neutral project for the Zhejiang Branch of China Construction Bank in 2013
24	Carbon neutral project for The Six China International Forum of Ecological Competitiveness 2014
25	Carbon neutral project for the APEC 2014
26	Carbon neutral project for 2014 Green Enterprise Alliance Environmental Summit
27	Carbon neutral project for the China Charity Fair 2014
28	Carbon neutral project for the 1st national marriage project in 2014
29	Carbon neutral project for the 2015 meeting of the heads from departments of forestry
30	Carbon neutral project for the Annual Summit of China Green Companies in 2015
31	Carbon neutral project for the meeting on "aviation town and drone application" by DFUAS in 2015

(continued)

Table 2 (continued)

No.	Project
32	Carbon neutral project for the 21st mountain hiking event organized by the Chaoer Forestry Bureau of Inner Mongolia and the 2nd China Green Carbon Sink Festival in 2015
33	Carbon neutral project for Sino-Finnish marriage projects in 2015
34	Carbon neutral project for the promotion and exchange event of Puer Forestry Culture in 2015
35	Carbon neutral project for the 1st forum on Dongjiangyuan afforestation and forestry sustainable management in 2015
36	Carbon neutral project for the thematic side event on "China's forestry in action: combating climate change" during the UN Climate Change Conference in Paris in 2015
37	Carbon neutral project for the side event on "developing carbon sink city to combat climate change" during the UN Climate Change Conference in Paris in 2015
38	Carbon neutral project for the Annual Summit of China Green Companies in 2016
39	Carbon neutral project for the First Zero-carbon Wedding at a tourist attraction
40	Carbon neutral project for the 3rd Tongshan series events of The Third China Green Carbon Sinks Festival in 2016
41	Carbon neutral project for The Third China Green Carbon Sinks Festival and the founding ceremony of carbon sink forest during the voluntary tree-planting day in Shenyang
42	Carbon neutral project for the forestry carbon sink training course in Daxing'anling of Inner Mongolia in 2016
43	Carbon neutral project for The Third China Green Carbon Sinks Festival and Ecological Restoration Display Activities
44	Carbon neutral project for the 5th International Forum on Clean Energy
45	Carbon neutral project for the Annual Summit of China Green Companies in 2017

come with higher costs, greater carbon emissions, and poorer results. For many reasons, it has been difficult to organize large-scale voluntary tree-planting activities in various locations in recent years. Therefore, there is an urgent need to innovate the approach for people to fulfill their tree-planting responsibility. For this reason, the Beijing Municipal government issued the Urban Greening Regulations in 2010, and will donate to "purchase" forest carbon as one of the 18 ways in which citizens can fulfill their obligations of planting trees. Drawing on Beijing's practice, the China Green Carbon Foundation launched the first "Afforestation of the Motherland Low-Carbon in Action" tree-planting festival in 2011, creating an online platform for the general public to implement "in-house and low-carbon tree plantation." Since 2011, on March 11 of each year, the China Green Carbon Foundation, in cooperation with various cities and institutions across the country, jointly initiated the launching ceremony of the "Afforestation of the Motherland, Low Carbon in Action" event. Citizens are guided to "plant trees without leaving home, fulfill obligations and offset carbon footprint." In order to facilitate people to adopt "forest carbon purchase" to fulfill the obligations of tree planting, the China Green Carbon Foundation has deployed voluntary tree-planting bases in more than 70 cities (counties) across the country. Citizens can choose their afforestation sites and number of plantation according to their own wishes by donating to "forest carbon purchase" of the China Green Carbon Foundation through online donation, bank transfer, post office remittance, and other methods. After successful donation, information such as donor name, planting site, number of trees planted, and the amount of carbon obtained will be publicized on the official website of the China Green Carbon Foundation (www.thjj.org), and citizens can download or print it from this website their "purchase certificate of forest carbon for the voluntary tree-planting project of the China Green Carbon Foundation." For example, in 2011, the Wenzhou Municipal Party Committee and government leaders completed voluntary tree planting in the conference room through online donation to "forest carbon purchase," eliminating the difficulty of forestry departments in finding tree-planting location and the cost of preparing afforestation tools. This approach not only improves the fulfillment of responsibility, but also ensures the quality of afforestation. The platform set up by the China Green Carbon Foundation has created conditions for many companies and institutions to voluntarily plant trees with 100% of fulfillment.

(IV) Promotion of forest carbon knowledge

Coping with climate change is a hot topic today. **Forest carbon** is an unfamiliar concept that is too professional and difficult to understand. Therefore, popularizing **forest carbon** knowledge and promoting carbon absorption and carbon sequestration of green plant are major contribution to combating climate change and important functions of the China Green Carbon Foundation. It is also one of the normal tasks of nationwide public fund foundation. As a result, the China Green Carbon Foundation has organized a variety of promotional and publicity campaign, which played a positive role in promoting the implementation of national policies and measures to combat climate change.

(1) Creating China green **forest carbon** Festival

From June 5 to June 25, 2014, the first "China Green **forest carbon** Festival, Green Charm-Bamboo Musical Instrument and Bamboo Culture and Art Exhibition" was held for a period of 20 days. It was jointly organized by the China Green Carbon Foundation, the National Centre for the Performing Arts and the Beijing Bamboo Orchestra, and co-sponsored by Anji County People's Government (known as the hometown of Chinese bamboo) and Lin'an Municipal Government in Zhejiang Province, Yong'an Municipal Government in Fujian Province, and Changning County People's Government in Sichuan Province. This is the first time that a public welfare performance assumes such great importance in fine arts.

The event was centered on bamboo. As pointed out by Zhang Yongli, Vice Administrator of the SFGA, at the opening ceremony, Bamboo is a critical part of forest resources and forest ecosystems, and it plays a key role in maintaining ecological balance, preventing soil erosion, and conserving water resources. It is showing great potential in bio-energy supply, biodiversity conservation, degraded land restoration, etc. China has become the world's No. 1 in terms of bamboo species, plantation acreage, bamboo product processing, export, and trade, and it is the well-known "kingdom of bamboo". The bamboo culture, especially the bamboo musical instruments, also embodies the profound traditional Chinese culture and becomes an integral part of the Chinese national art treasures. It calls for us all in vigorously promoting and passing on the inheritance.

The First "China Green **forest carbon** Festival, Green Charm-Bamboo Musical Instrument and Bamboo Culture and Art Exhibition" featured more than 200 kinds of bamboo-made musical instruments around world and bamboo handicrafts from the "Hometown of Bamboo" in China. "The Sound of Bamboo Breeze—Theme Concert on the World Environment Day" was organized with all instruments made from bamboo. The audience enjoyed the heavenly sound of naturally grown bamboo. In addition, in order to promote public understanding of bamboo and bamboo music, forest carbon and the special functions and roles of forestry in combating climate change and protecting the environment, the festival also held lectures on carbon sink, bamboo calligraphy and painting, bamboo music, and low-carbon environmental protection. These physical objects, photos, videos, and lectures, while spreading the knowledge of Chinese bamboo culture to the audience, highlighted the special status and important role of forest carbons in combating climate change. The festival was listed as one of the major events of the World Environment Day for that year by the UNEP Beijing Office.

(2) Multiple forms of publicity in carbon sink

To further popularize the knowledge of forest carbon and guide the public to practice low-carbon life through afforestation and tree planting, the China Green Carbon Foundation designed and issued the world's first **forest carbon** gift card. The set of cards includes Spring Festival greeting cards, Christmas cards, teacher's day greeting cards, Valentine's Day cards, and so on. A variety of **forest carbon** cards can also be designed according to public needs, such as adult cards, golden wedding commemorative cards, etc.

In addition, with the support of National Publication Foundation, the China Green Carbon Foundation edited and published the "China **forest carbon**" book series, such as *Forestry Carbon Sequestration in China, Carbon Inventory Methods, Basic Knowledge of Forestry Carbon Sequestration, Forestry Carbon Property Right in China, Forestry Carbon Management in China: Exploration and Practices, Methodologies for Forest Carbon Sequestration Projects, A Forestry Case of China Certified Emission Reduction Project,* and *Theory and Practice of Green Carbon Communication*, as well as other books promoting green **forest carbon**. The High School Affiliated to Beijing International Studies University co-authored four Sino-English bilingual school textbooks including *Forest carbon and Climate Change*, *Chinese Wildlife Conservation, China Desertification Prevention and control,* and *Wetlands in China*. The school allocates dedicated hours for this purpose for students in Junior 3 and Senior 1. It broadened the horizon of middle school students and laid the foundation for cultivating a new generation with international perspective and ecological protection awareness.

3 Voluntary FCCB Projects

In July 2010, FAW-Volkswagen Audi sales division launched the "Audi Low Carbon Action" project. This project is a long-term environmental protection project that advocates low-carbon emission reduction. The "Green Fund" was established under the auspices of the China Biodiversity Conservation and Green Development (CBCGDF). The first project supported by the fund is "Audi Forest for carbon—panda rescue in action at north latitude 28°." The project plans to plant 2,000 mus of forest in the Shenguozhuang Provincial Nature Reserve in Yuexi County, Liangshan Prefecture, Sichuan Province in 2011, which will restore the corridors between giant panda populations in the natural reserve, and protect and rebuild the habitats for giant panda. It is estimated that the project will absorb 40,000 tons of carbon dioxide within 30 years to offset part of the carbon emission generated by FAW, Volkswagen Factory, AviChina, and car owners who are keen on environmental protection. The project also provided job opportunities and skills training for local communities to improve the lives of local ethnic minorities.

From 2012 to 2014, FAW-Volkswagen Audi Sales Division continued to cooperate with the Beijing Shanshui Conservation Center and adopted China's first voluntary carbon certification standard in the Baicaopo Nature Reserve in Jinyang County and the Yele Nature Reserve in Mianning County of Liangshan Prefecture in Sichuan Province, namely, the "Panda Standard" for project development and development of multi-benefit forest restorations for giant panda habitats. The resulting carbon dioxide emission reduction will be used to offset the carbon emission from production and business operation, and to disseminate to the public the concept of credible and transparent carbon mission reduction. The plans to complete the recovery of 5,000 mus of FCCB Project in 3 years, including 1,500 mus in 2012, 1,500 mus in 2013, and 2000 mus in 2014.

3.1 *Project Distribution*

The total land area of the project from 2011 to 2013 is 5,000 mus. It is distributed in three counties of Yuexi, Jinyang, and Mianning. It involves three natural reserves in Shenguozhuang, Yele, and Baicaopo, covering 455 households of 1468 people. All of them are from the Yi minority group, with the number of people living in poverty accounting for over 90%.

3.2 *Project Activities*

In order to achieve the goal of FCCB project, the project carried out a series of activities, as shown in Table 3.

Table 3 Activities of FCCB project

Year	Planned activities	Completion status
2011	Siting and feasibility research	Complete siting and feasibility analysis
	Afforestation design and activity review and approval	Complete afforestation design, the activity is reviewed and approved through the Shenguozhuang Natural Reserve
	Seedling purchase, soil preparation, plantation, and tending	Complete seedling purchase, soil preparation, and 2,000 mus of plantation and seedling tending
	Tending and plantation	Conduct plantation and seedling tending for 2 times
	management and control	Work on tending the seedlings
	Supervision of afforestation and outcome inspection	Onsite supervision of afforestation, conduct outcome inspection for 3 times
2012	Develop project development plan	Complete drafting the project investigation guidance and formulate development plan
	Select project site	Complete site selection and initially identify project area
	Project baseline and additionality analysis	Complete project baseline and additionality investigation and data compilation analysis
	Project planning	Complete project planning
	Draft PIN	Complete PIN draft
	Afforestation and forest management design	Complete afforestation design
	Plant 1,500 mus of forest	Complete 1,500 mus of afforestation
	seedlings tending	Work on seedling tending
	Seedlings management	Conduct seedling management
	Outcome evaluation	Complete project evaluation
	Technical training and guidance	Develop technical training and guide project implementation on site
	Progress report	Already submitted once
2013	Boundary and land eligibility assessment	Confirm the boundary
	Project baseline investigation	Complete project baseline investigation and related reports
	Additionality investigation	
	Afforestation operation design	
	Draft monitoring plan	
	PDD	Complete PDD draft

(continued)

Table 3 (continued)

Year	Planned activities	Completion status
	Project application, verification, and certification	
	Afforestation and forestry operation design	Complete afforestation design
	Plant 1,500 mus of forest	Complete 1,096 mus of afforestation at Yele
	Seedling tending	Work on seedling tending
	Seedling management	Conduct seedling management
	Outcome evaluation	Complete project evaluation
	Technical training and guidance	Develop technical training and guide project implementation on site

3.3 Project Output and Impact

Audi's 2011 annual project is located in Masizulai, Laji Village, in the northern part of Laji Township, which is inside the Shenguozhuang Nature Reserve in Yuexi County of Liangshan Prefecture in Sichuan Province. The project plans to plant an area of 2,000 mus and actually completed an area of 2,000 mus, including 620 mus of fir, 1060 mus of *Pinus armandii,* and 320 mus of *Populus szechuanica*. It is estimated that 40,875 tons of CO_2 equivalent will be generated within 30 years. The afforestation will increase the forest area by 2,000 mus for the nature reserve and increase the forest volume stock by 21,765 m^3, which will promote the restoration of the habitat for giant pandas in Shenguozhuang Nature Reserve and increase the income for the minority communities in poverty.

Audi's forest carbon project for year 2012 is located in Yele Nature Reserve in Mianning County and Baicaopo Nature Reserve in Jinyang County, Liangshan Prefecture, Sichuan Province. The project plans to plant an area of 1,500 mus, including 1,000 mus at Baicaopo and 500 mus of Yele. The project actually completed an area of 1,500 mus, of which Baicaopo accounts for 1,000 mus, including 500 mus of fir, 500 mus of weeping cedar; Yele accounts for 500 mus, including 212 mus of Chinese spruce, 288 mus of *Alnus cremas-togyne*. The project will be developed in accordance with the Panda Standard and it is estimated that the project will generate 14,983 tons of CO_2 equivalent emission reduction within 30 years. The project afforestation will increase the forest area by 1,500 mus for the nature reserve and increase the forest volume by 8,777 m^3, which will promote the restoration of habitats for rare species in Yele Nature Reserve and Baicaopo Nature Reserve, and increase the income for poor minority communities.

4 Online Public Forest Carbon Projects

4.1 Project Intent

After the adoption of the Paris Agreement in 2015, Chinese Internet companies have adopted green low carbon as a means of connecting with the public so as to fulfill their social responsibilities. In August 2016, the Ant Financial, a leading Internet company in China, launched the Ant Forest Program. The company developed an ant forest application on the Alipay APP with nearly billion users. The application quantifies users' low-carbon behavior, including the offline support shown in Fig. 9, train and shared bicycle travel, green package, and other behaviors. Using the carbon measurement method provided by the Beijing Environment Exchange, these practices are automatically converted into energy by Alipay and deposited into users' virtual carbon account. When the carbon (energy) reaches certain amount, Ant Financial will work with public welfare organizations to plant a real tree for users to offset their carbon footprint (Fig. 10).

After the Ant Forest came online, it received great attention from users. After 1 year, the number of users of the application reached 230 million. It is estimated that, through green and low-carbon activities, users have reduced more than 122 tons of carbon dioxide emission. At the same time, Ant Financial worked with SEE Foundation, China Green Foundation, and China Green Carbon Foundation to have planted more than 10 million trees for users. In September 2017, Ant Forest announced a plan of launching a public platform. It worked with more public social organizations to incorporate forest protection into ant forest project, and announced that it would invest RMB 200 million in the plantation and management of ant forests.

4.2 Ant Forest and Public Carbon Sink

Ant Forest is an innovative approach of connecting the public and carbon trading with mobile Internet tools. The innovation of the project lies in the application of cloud computing and big data, which records low-carbon behavior of the public and encourages the public to carry on with low-carbon life through afforestation. This is a typical forest carbon project for public interest. Carbon neutral approach is used to encourage the public to practice low-carbon activity.

It should be pointed out that the development of ant forest carbon forest is not a forest carbon in its strict sense. Ant forest and partners, while planting trees, adopted CCB standards to emphasize forest's positive role in climate change, biodiversity, combating desertification, and community development. The technical regulations for afforestation are based on the afforestation procedures formulated by the State Forestry and Grassland Administration.

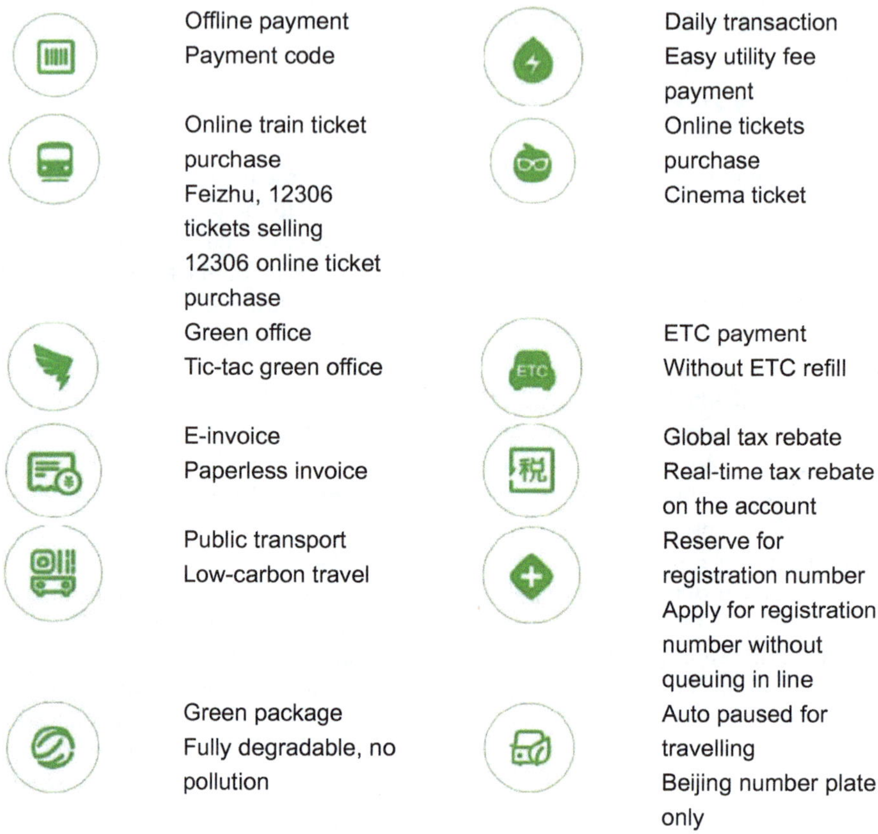

Fig. 9 Low-carbon behaviors accountable for ant forest

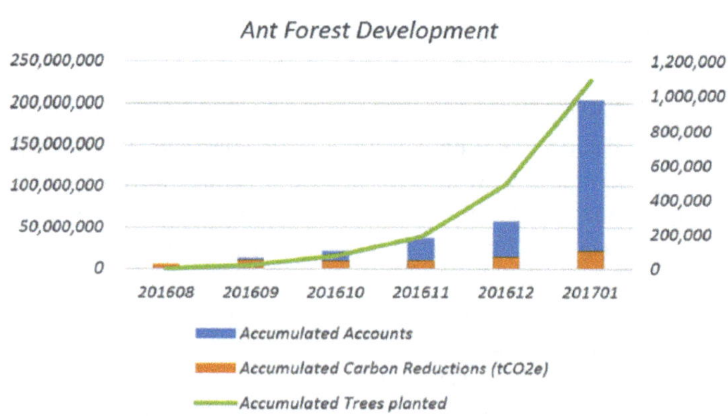

Fig. 10 User increase reduction quantity from low-carbon practice and plantation quantity of ant forest (data from public source)

Because of its charitable and voluntary nature, Ant Forest also introduced methods such as the management in natural reserves that can also enhance forest carbon. Its publicity significance is far greater than the carbon effect brought by the project itself.

It can be predicted that, in the context of low-carbon policies in China and the establishment of carbon market in China, the future of ant forest has great potential to change from a purely voluntary nonprofit carbon project to a real carbon account. But whether this can be realized depends on the development of China's carbon market in the future, and on how the application can be iteratively adapted to China's low-carbon policy and user preference. Whatever, the ant forest has received investment over RMB 100 million from Ant Financial and netizens to support afforestation since its launch. This has produced considerable social benefits.

4.3 Experience, Lessons, and Future of Public Voluntary Forest Carbon Project

Since the establishment of the China Green Carbon Foundation, the public voluntary forest carbon project has been fully funded, with sound contract management, strict supervision, and measurement science. The project is well implemented and the expected goal is achieved. The project not only has the output of emission reduction, but also comes with multiple benefits.

4.4 Experience in Implementing Public Voluntary Forest Carbon Project

(1) A series of effective project implementation management models were created

Based on the principle of seeking truth from facts and adapting to local conditions, on the basis of strengthening contract management, a variety of project implementation and management models were created featuring forestry authorities at all levels exercise on the spot supervision, and special funds, state-owned forest farms, and afforestation companies engaged in organization and implementation. The supervision is in place, the project is executed in an orderly manner, and the development is effectively managed.

(2) Spread the concept of forest and play a demonstrative role

Based on the special requirements of afforestation technology on land eligibility, soil preparation, and carbon measurement, the design, implementation, and management of afforestation projects for carbon are different from conventional state-subsidized afforestation. In particular, in some arid and semi-arid regions, barren hills and mountains that used to be treeless for many years have began to show lush young forests. Monuments and project descriptions of "forest carbon" were erected beside the afforestation plots, which is a unique carrier for promoting green forest carbon. More and more enterprises and the public are paying attention to the public voluntary forest carbon project. They participate in carbon offsets and eliminating carbon footprint through "forest carbon purchase" and practices "low carbon production and low carbon life." The public voluntary carbon project has become an innovative carrier for companies to fulfill their social responsibilities and display corporate image.

(3) Improve the ecological environment and increase farmers' income

The successful implementation of the project has played a positive role in improving the local ecological environment and promoting farmers' employment and income. For example, in 2010, the China Green Carbon Foundation invested RMB 3.75 million to build 5,000 mus of forests in the counties of Xiangyuan, Xiyang, and Pingshun, and Shanxi Province. It is planned to spend 10 years in absorbing all the carbon emission generated from the UN Climate Change Conference in Tianjin. During the operation of the project, farmers in the project area are expected to receive income of RMB 2.6 million and forest by-product and timber income equivalent to about RMB 7 million.

4.5 Problems in the Implementation of Public Voluntary Forest Carbon Project

After certain stages of development, there are still some problems with public voluntary forest carbon projects as follows:

1. The cost of forest carbon is underestimated. Ensuring the healthy growth of forests and obtaining the expected forest carbon require not only adequate investment of fund and manpower in afforestation, but also the investment of same resource after the afforestation to ensure the benefits of forest carbon. In addition, from the perspective of biodiversity, the project requires the deployment of local tree species as much as possible. Therefore, the cost of afforestation is higher than that for singly planted trees. What is more, the cost of some forest carbon projects is mainly afforestation cost, resulting in the misunderstanding that the cost of public voluntary forest carbon project is lower than that of obligatory forest carbon project in the market.
2. The enthusiasm of enterprises in voluntary emission reduction and donation to afforestation need further improvement. At present, low-carbon transition and green development have become the consensus of the society. More and more people are beginning to donate to "forest carbon purchase" for their low-carbon lives. Some companies also shift the focus of donating funding to fund forest carbon project. However, there are not enough enterprises or large emitters in key emission areas that are involved in voluntary emission reduction and fund for the afforestation project for carbon.
3. The dissemination and publicity on forest carbon are insufficient. Forest carbon is a new concept. The forest carbon project is a new thing. Due to the lack of comprehensive and systematic publicity, it is not widely accepted and understood by the public With the launch of China's carbon trading pilot project and

the establishment of a unified national carbon market, the forest carbon projects for market transactions will be developed in strict accordance with the relevant national policies and CCER forestry carbon sink project methodology, while the forest carbon generated from public voluntary forest carbon projects will continue to function in voluntary emission reduction due to their roles in ecological protection and community development. They serve as a supplement to national mandatory emission reduction and require further publicity and advocacy, so that more public funds will enter into areas that are difficult to carry out in market mechanism but have better social benefits for the development of public voluntary forest carbon projects.

In response to existing problems, organizations committed to public voluntary forest carbon projects should develop, design, and use forest carbon for responsible corporate units to implement voluntary emission reduction and fulfill their social responsibility. Companies can plan ahead and actively respond to global low-carbon development. This will prepare for the companies for sustainable development, ecological civilization, environmental protection, and coordinated development of climate change. At the same time, various measures must be adopted to disseminate forest carbon concept to the entire society, vigorously publicize the comprehensive effects of forest carbon projects, and respond to the special functions and roles of climate change. The whole society will be mobilized to care for, support, and participate in forest carbon projects.

4.6 Prospects of Public Voluntary Forest Carbon Projects

With the implementation of China's low-carbon transition and green development strategy, combined with the implementation of the concept of ecological civilization and the promotion and publicity of forest carbon, multiple value forest carbon project will gain the recognition, acceptance, and support of more companies and the public.

References

China Green Carbon Foundation (2011a) Validation and verification manual for forestry-based carbon sequestration project in China (Trial)
China Green Carbon Foundation (2011b) Provisional management method for registering forestry carbon sink projects of China green carbon foundation
SFGA (2010) Technical regulations on carbon sink afforestation (Trial) (BZZ [2010] No. 84)
SFGA (2011a) Measurement and monitoring guidance for carbon sink afforestation project (BZZ [2011] No. 18)
SFGA (2011b) Provisional method for measuring and monitoring forestry carbon sink (BZZ [2011] No. 18)

Potential of Forest Management Carbon in China

Wen Zhang and Caifu Tang

Summary

Although forest management carbon project is not a qualified CDM project under the Kyoto Protocol, forest management carbon project has developed rapidly in the voluntary carbon market. In the past few years, forest management carbon trading quantity and trading volume have increased exponentially. China's forestry has made remarkable achievement in the past 30 years, and forest area and volume have increased substantially, but due to deforestation, unreasonable selection of tree species, insufficient investment in A/R, the seedlings and A/R quality are poor. There is no effective tending and management, plus frequent natural disasters, the quality of Chinese forest, especially plantations, and collectively and individually owned forests are worse. There are also large areas of low-function forests, and the economic value and ecological service function of the stand are both left much to be desired. In recent years, the country has initiated and gradually increased efforts in forest tending and restructuring low-function forest, and explored a fiscal subsidy policy for forest tending. In the "Outline for National A/R and Greening Plan (2010–2020)," ambitious quantitative targets are set for forest tending and low-function forests.

There are very few studies on carbon benefits of transforming China's low-function forest. Most people think that carbon sink has huge potential. Through our case study of transforming and planning for the low-function cypress plantation in the middle of the Sichuan Basin and the low-function forest in southeastern Yunnan, it is shown that the presence or size of the carbon benefit of transforming low-function forest depends on the existing stand, low-function transformation measures, and the management purposes in relation to the baseline scenario of maintaining the status

---

W. Zhang (✉)
The Nature Conservancy China Program, Kunming, China
e-mail: 1831841301@qq.com

C. Tang
Sichuan Green Carbon Ltd., Chengdu, China
e-mail: 631592345@qq.com

Z. Lu et al. (eds.), *Forest Carbon Practices and Low Carbon Development in China*, https://doi.org/10.1007/978-981-13-7364-0_8

quo. But in general, carbon benefits are much lower than A/R projects. For plantation forests with excessive forest density due to lack of tending, if the target is on long-cycle and large-diameter product, namely, there is no main cutting during the crediting period, there will be obvious carbon benefits. However, if the target is on short-rotation industrial raw material or fast-growing high-yield forest, namely, one or more main cuttings occur during the crediting period, the carbon benefits are very limited, and even in comparison with baseline scenario, the long-term carbon reserve in the biomass will decrease. For example, in structural adjustment and species replacement, if the adjusted or replaced species are operated in short rotational period, then it is likely to be net carbon sources compared to baseline scenario. Therefore, if the low-function forest is transformed to produce net carbon benefit, or even have it included in carbon trading, short-term rotation should be avoided as much as possible, and long-cycle and large-diameter products should be cultivated. If main cutting is required, selective cutting should be adopted as much as possible to increase the long-term average of carbon reserve.

1 Background

1.1 Overview of Carbon Project and Methodology for International Forest Sustainable Management

During the first commitment period of the Kyoto Protocol (2008–2012), increasing carbon through afforestation/reforestation and forest management has become one of the main solutions for industrialized countries to meet their greenhouse gas emission reduction targets under the Kyoto Protocol. According to preliminary estimates, more than half of among the 24 industrialized countries that choose to use forest carbon during the first commitment period of the Kyoto Protocol, more than half can achieve offset by carbon from forest management (Zhang 2011). In September 2009, at the opening ceremony of the United Nations Climate Change Summit, President Hu Jintao proposed to enlarge forest resources and increase forest carbon. By 2020, China's forest area will increase by 40 million hectares compared with 2005, and forest volume will increase by 1.3 billion cubic meters. The increase in forest volume (i.e., the increase in carbon sink) is mainly from forest management.

However, during the first commitment period, forest management was not included as qualified CDM project activities. Based on the understanding that forest management can enhance forest carbon, in the voluntary market, forest management projects are recognized by multiple voluntary carbon standards, such as CCX, VCS, CCB, CFS, China's CCER, and Panda standards. In particular, since 2008, the carbon trading volume of forest management projects in the voluntary market has more than doubled, increasing from 430,000 tons of carbon dioxide equivalent in 2008 to about 3 million tons of carbon dioxide equivalent in 2010, and its share in the voluntary carbon market is up from 1 to 5% (Peter-Stanley et al. 2011). Since 2010, VCS has approved the methodology of five forest management projects, four of which are suitable for protecting forests that have been degraded due to overharvesting. China has fewer situations in this regard, so its relevance to China is low. In addition, in order to extend the rotational period of commercial forests and increase Carbon Stock, it is suitable for China to change from forest management that aims to cultivate medium and small-diameter forests to that with the goal of cultivating large-diameter timber (Table 1).

1.2 Overview of China Forest Quality

According to the seventh national forest resource inventory, the national forest area is 195.4522 million hectares, the forest coverage is 20.36%; the total standing volume is 14,913 billion cubic meters, and the forest stock volume is 13,721 billion cubic meters. Among these, the plantation area is 61.6884 million hectares, the volume

Table 1 Methodologies of forest management (http://v-c-s.org/methodologies/find-a-methodology?Title=&tid=14&=Search)

No.	Methodology	Conditions	Carbon reserve	Emission source	Leakage
VM0003	Methodology of extending the rotational logging period	• Both the baseline scenario and project scenario involve clear-cutting or block clear-cutting (less than 1 ha) • The forest must pass the FSC certification or pass the certification within 1 year after the project starts • The project period must be determined • The project does not include peat forests, and the project does not change the proportion of wetlands • Management measures for project scenarios and baseline scenarios must be determined • If fire is involved in the forest management process, it should be ensured that it does not spread beyond the project boundary (causing biomass burning outside the boundary) • The project start date is the same as the start date of crediting period	Aboveground biomass, below-ground biomass, deadwood, and harvest wood product	Biomass burn-ing	Market leakage

(continued)

Table 1 (continued)

No.	Methodology	Conditions	Carbon reserve	Emission source	Leakage
VM0005	Methodology of transforming low-yield forest to high-yield forest	• The project site must be an overharvested, degraded tropical evergreen natural rainforest • The project aims to prevent re-harvesting the existing forests or to restore the degraded tropical evergreen natural rainforests by removing climbing plants, thinning, or replanting • In the baseline scenario, the forest will not return to normal growth • At the beginning, the soil carbon pool is in a stable state, or the soil carbon will increased more or reduce less under the project scenario than the baseline scenario • Site preparation will not lead to a significant reduction in soil carbon during the project period • No nitrogen fertilizer is used in the project activities • There is no main cutting in the project scenario during the crediting period • Under the project scenario and the baseline scenario, no biomass burning, fuelwood collection, removal of dead wood and litter are involved in the project boundary • When using the control area to determine the parameters related to baseline scenario, the coverage of the control area is greater than or equal to 75% of that for the project area • There is no flood irrigation or drainage measures under the project scenario	Aboveground biomass, belowground biomass (optional), deadwood, and harvest wood product	Fossil fuel burning	Market leakage

(continued)

Table 1 (continued)

No.	Methodology	Conditions	Carbon reserve	Emission source	Leakage
VM0010	Protect to-be-logged forest	• Wood harvesting is planned under the baseline scenario • Under the project scenario, the use of forests does not involve commercial harvesting and activities leading to forest degradation • The wood to be harvested must be determined by forest survey • The project has clear boundaries • The baseline scenario does not include updating to plantation forest after harvesting • The project site does not include wetlands and peatlands	Aboveground biomass, deadwood, and harvest wood product	Biomass burn-ing	Activity transfer and market leakage
VM0011	Protect logged and to-be-logged forest	• The project scenario does not involve harvesting and biomass burning • At least 10 years before the start of the project, there should be undisturbed forest land, or forest that has degraded due to harvesting • Forests are tropical forests • Harvest wood products are sawn timber, pulp, or commercial firewood • Degradation drivers are commercial logging permitted by government or law • Harvesting in the baseline scenario is selective cutting	Aboveground biomass, deadwood, and harvest wood product	Fossil fuel burn-ing	Activity transfer and market leakage
VM0012	Temperate and northern forest man-agement	• The project is to protect forests that are to be harvested or degraded due to harvesting • The project is located in the temperate zone or the northern forest belt • The harvesting level does not exceed 5% of annual logging of the baseline scenario • The project site is not a peat forest • The project does not cause changes in the wetland area within the project boundary	Aboveground biomass, below-ground biomass, deadwood, and harvest wood product	Fossil fuel burn-ing	No activity transfer leakage

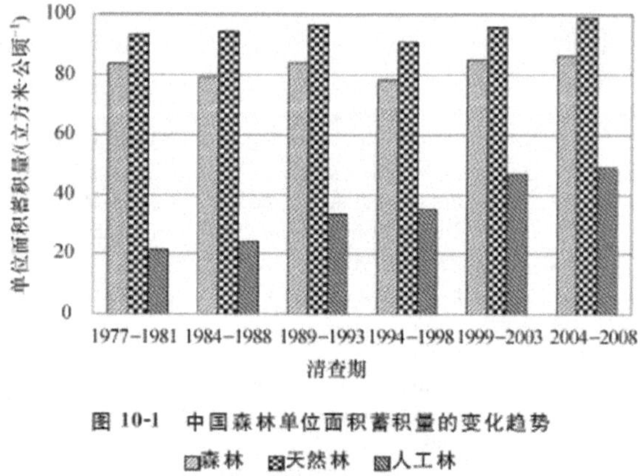

单位面积蓄积量/（立方米·公顷$^{-1}$）	volume per unit area/(cubic meter·hectare^{-1})
清查期	Inventory period
森林	Forest
天然林	Natural forest
人工林	Plantation

Fig. 1 Trend of per unit forest volume in China. From inventory period (1989–1993) to inventory period (1994–1998), the per unit volume is on the decrease. This is mainly because the threshold of forest canopy density is set at lower value. From inventory period (1994–1998), the defined canopy density is adjusted from ≥0.3 to ≥0.2.)

of plantation is 1,961 billion cubic meters, and the plantation area ranks the first in the world (SFGA 2009). The quality of forest resources in China has continued to increase. The volume per unit area has increased from 78.06 m^3/ha in the inventory period (1994–1998) to 85.88 m^3/ha in the inventory period (2004–2008), an increase of about 10%. In particular, the plantation increased from 21.48 m^3/ha in the inventory period (1977–1981) to 49.01 m^3/ha in the inventory period (2004–2008), an increase of 128.2% (Fig. 1). The ratio of plantation per unit area to the volume of natural forest unit area increased from 0.23 in the inventory period (1977–1981) to the recent 0.50.

Despite this, the quality of forests in China is still poor, which is mainly in plantation and collectively owned forest. Currently, the area of plantation per unit is only half of that of natural forests (Fig. 1), especially for the middle-aged forest, the difference between plantation and natural forest is increasing (Fig. 2). In addition to the different standards of age classification for plantation and natural forest, it is mainly affected by the following aspects: insufficient investment in A/R in China, poor quality of seedlings and A/R, ineffective tending and management of plantation, and vulnerability of plantation to natural disasters (forest fires, pests and diseases, snowstorms, etc.).

From the perspective of ownership, China's state-owned forests are of higher quality, while collectively and individually owned forests are poor. Although the

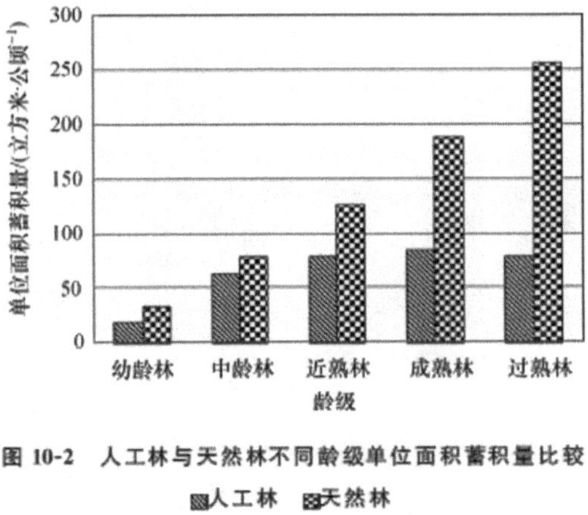

单位面积蓄积量/（立方米·公顷-1）	Volume per unit area/(cubic meter·hectare^{-1})
幼龄林	Young forest
中龄林	Middle-aged forest
近熟林	close to mature forest
过熟林	Over-mature forest
龄级	Age class
天然林	Natural forest
人工林	plantation

Fig. 2 Comparison of volume per unit for different age classifications between plantation forest and natural forest

unit area of volume for collectively and individually owned forests is increasing, it is still very low (Fig. 3). At present, the unit area of stock (66.3 m^3/ha and 44.9 m^3/ha) is only 48.1% and 35.9% (SFGA 2009) of state-owned forests, respectively. Even in terms of origins and ownerships, the quality of state-owned plantation is still poor and the unit area of stock is less than half of the state-owned natural forests; the collectively owned plantation is worse than the state-owned plantation; the collectively owned natural forest is only 40% of state-owned natural forest in terms of unit area of stock (Fig. 4).

Therefore, from the perspective of per unit area of stock, the poor quality of forests in China has mainly reflected in plantation and collectively owned natural forests, especially collective and individually owned plantations.

The quality of forest resources is also reflected in its structure, health, and ecological service functions. The arbor forest in China enjoys 58.02% intact structure. Among arbor forests, pure forests accounted for 62.59%, mixed forests accounted for 37.41%, while among plantation forests, pure forests accounted for 86.8%. Compared to the 1999–2004 inventory period, the proportion of mixed forests is increased by 9.17%. The area at health level accounted for 72.33%, while those in subhealth, medium health, and unhealthy levels accounted for 21.44%, 4.72%, and 1.51%,

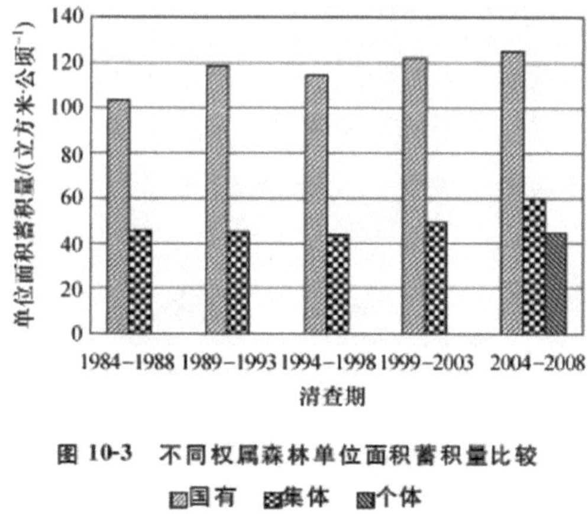

单位面积蓄积量/（立方米·公顷-1）	Stock per unit area/(cubic meter·hectare^{-1})
清查期	Inventory period
国有	State-owned
集体	Collectively-owned
个人	Individually-owned

Fig. 3 Comparison of unit area of volume of forests with difference ownerships

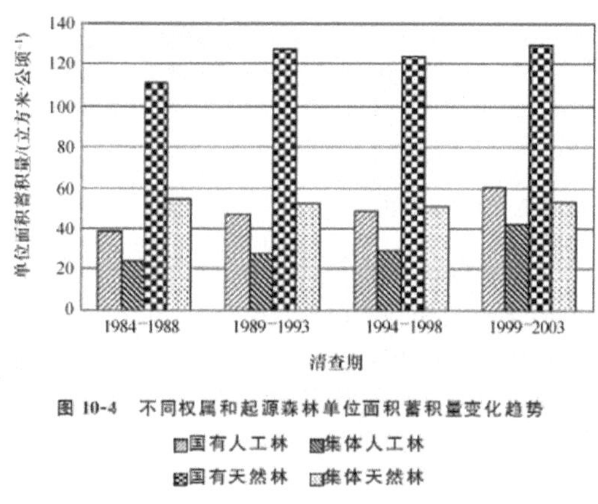

单位面积蓄积量/（立方米·公顷-1）	Stock per unit area/(cubic meter·hectare^{-1})
清查期	Inventory period
国有人工林	State-owned forest
集体人工林	Collectively-owned plantation
国有天然林	State-owned natural forest
集体天然林	Collectively-owned natural forest

Fig. 4 Trend of unit area of volume of forest with different ownership and origin

respectively. According to the comprehensive assessment, the area with good medium and poor quality of arbor forest accounted for 16.66%, 60.96%, and 22.38%, respectively, with the overall level at medium range. In terms of ecological service function, the areas with good, medium, and poor ecological function levels are 11.31%, 79.74%, and 8.95%, respectively (SFGA 2009).

Due to poor quality and health status, the forest is weak in withstanding natural disasters, which in turn affects the quality of forest resources. According to the seventh forest resource inventory, the affected area of China's arbor forest is 17.6174 million hectares, accounting for 11.32% of the total. Among these, the area affected by pests and diseases is 8.9302 million hectares, accounting for 50.69% of the total; the area affected by fire is 3.0646 million hectares, accounting for 17.39% of the total; the area affected by other natural disasters such as climate is 5.6232 million hectares, accounting for 31.92% of the total (SFGA 2009). For example, in early 2008, southern China suffered extremely severe coldness, heavy rain, snow, and ice disasters. According to preliminary assessment, the affected area of forest in 19 provinces (autonomous regions and municipalities) was 313 million mus. Among these, the affected area of forest is 290 million mus, the affected area of bamboo is 44.5 million mus, the affected area of growing forest is 205.6 billion mus, the affected area of nursery is 2.15 million mus; the loss of forest stock is 371 million cubic meters and that of bamboo is 2979 million; the affected area of economic forest is 27.48 million mus and the seedlings are 10 billion. The direct loss of forest resources amounted to 58.2 billion RMB, and the loss of forestry is far reaching, and its recovery is also a long-term process. Among the affected forests, plantation, young- and middle-aged forests, bamboo forest, and exotic tree species suffer more than natural forest, mature forest, and native tree species. For example, wetland pine, eucalyptus, and bamboo forest are almost 100% damaged, and the losses are severe; pure forests suffer more than mixed forests (Zhang and Shuhong 2010).

From September 2009 to spring 2010, the rainfall in most parts of Yunnan, Guizhou, Guangxi, Sichuan, and Chongqing was 50% less than the same period of previous years. In some areas, it was 70–90% less, and in southwest China, the precipitation was the lowest since the meteorological record in 1951, while the temperature is generally higher during the same period. For example, the average temperature in the whole winter of Yunnan has reached the highest level since 1950. High temperature and low rainfall have caused the most serious drought for the past 60 years, and a large number of forests including large numbers of original forest vegetation have died (Zhang and Shuhong 2010). In the spring and summer of 2011, the southwest region suffered another severe drought.

According to reports, nearly 5 million hectares of forestland in the country has been degraded to sparse forests, and more than 50 million hectares of forest land has been degraded to inefficient forest land with a canopy density of less than 0.4. The case is especially severe in the western and southern regions and the collectively owned forest areas (SFGA 2010a).

2 Overview of Policies on Forest Management and Its Implementation in China

Forest management is a general term for the scientific management of existing forests to maintain and improve the production and quality of forests, including controlling the formation, composition, structure, growth, and harvesting of forests. Sustainable forest management is to maintain the health and vitality of forest ecosystems through scientific and rational management of real and potential forest ecosystems, so as to maintain biodiversity and its ecological functions, and meet the demand for harvest wood products and their environmental service functions forests in the process of social and economic developments, and ensures and promotes the coordinated and sustainable development of population, resources, environment, society, and economy. Due to the fact that China has long been emphasizing on A/R over management, most forests in China lack management, and due to various other reasons, there are large scale but low-function forests. Therefore, from policy perspective, in recent years, the state aims to improve forest management and forest quality by strengthening forest tending and reforming low-function forest. Forest tending and low-function forest transformation are different and interconnected. The lack of tending is one of the causes of low-function forest. Forest tending is not equivalent to low-function forest transformation, yet it is also one of the measures for reforming low-function forest.

2.1 The Concept and Cause of Low-Function Forest

1. **The concept of low-function forest**

For a long time, due to various reasons, China has produced a large scale of low-quality forests, namely, low-function forests. The so-called low-function forest refers to the direct effect of human factors or the influence of nature-induced factors, which caused the structure and stability of forest stands to lose balance, the exhaustion of forest growth, and the degradation or loss of systematic function, resulting in significantly lower forest ecological function, forest product yield, or biomass than the average level of the same forest under the same site conditions. According to different origins, low-function forests can be divided into inefficient secondary forests and low-function plantation; according to different business objectives, low-function forests can be divided into low-function shelter forests and low-quality low-yield forests.

Low-function secondary forest refers to the low-function forest formed by original forest or natural secondary forest due to long-term human damage, and is divided into residual forest and inferior forest.

The residual forest refers to the forest stand that suffers disturbance, forest phase failure, structural imbalance, low canopy density, and serious soil erosion, with low economic value and ecological function.

The inferior forest refers to the stand of depleted quality seeds due to unreasonable utilization, the remaining species are of poor genetic quality, degraded natural development, thus losing its management value.

Low-function plantation forest refers to forests created by methods such as plantation A/R or human-induced regeneration, but become low-function forests due to improper A/R measures (such as improper selection of tree species or provenance, serious damage by pests, improper management measures, or poor management).

Compared to the same site conditions and the average value of stand under the same operating level (SFGA 2007), low-function forests need to meet one of the following conditions:

Forest stand with disastrous forest phase, low function, and results in degradation of forest ecosystems;

The quality stand seeds and species are depleting, and the number of quality forest trees with natural breeding ability is less than 30 plants per hectare;

Stands with a growth rate or biomass that is 30% lower than the average level of similar site conditions;

Stands of forests with canopy density less than 0.3;

Suffering from serious pests, droughts, floods, and natural disasters such as wind, snow, fire, etc., the proportion of dead wood (including dying wood) accounts for more than 20% of the stand per unit area (forest belt);

Fuel wood forest that has passed two times of logging and with a declining budding ability;

Bamboo forest that has low bamboo shoot rate or low bamboo yield due to excessive cutting, death and rotten bamboo/whip bamboo, and old bamboo whipping and cramming in the forest land;

Low-function forests due to unsuitable trees on unsuitable land or unreasonable provenance;

In addition to the abovementioned general standards, ecological standards and economic standards for determining low-yield forests have been developed (SFGA 2007). The specific judgments of these standards vary according to different forest species, business purposes, different climate types, and tree species.

2. **Causes of low-function forest**

The main reasons for low-function forests are as follows (Deng et al. 2010):

Site conditions are too poor;

Human disturbance: residual forest due to deforestation, excessive pruning, and logging;

Plantation forest violates the principle of suitable land for trees, resulting in poor quality of A/R and improper management: improper selection of tree species, extensive A/R measures, poor seedling quality, and tending failing to catch up in the wake of A/R. This causes poor survival and preservation rates, and replantation is not in time;

The stand density is too large: the young- and middle-aged forests could not catch up, or the density is too large, resulting in excessive forest stand, low growth, and weak ecological functions;

Natural disturbance: pests and diseases, fire, ice, and snow lead to low growth and ecological function.

2.2 Overview of Policies and Implementation Related to Forest Tending and Low-Function Forest Restructuring

Since the benefits of forest tending are often lower than the tending cost, and the state and local governments do not have the corresponding investment, quite a lot of low-function forests in China are caused by insufficient tending due to lack of funds. Based on this understanding, in recent years, the SFGA has gradually increased efforts in forest tending and low-function forest restructuring, and explored a fiscal subsidy policy for forest tending. According to the Central Work Conference on Forestry, "carry out pilot project for A/R seedlings and forest tending subsidies, the central government subsidies A/R of high-quality seedlings, young and middle-aged forests and low-yield forests, and gradually expand the scope of the pilot," on January 12, 2010, the SFGA held a video conference to officially launch the national pilot project of forest tending. The Ministry of Finance (MOF) and the SFGA decided to start pilot subsidies in 11 provinces and regions and Daxing'anling Group companies starting from the end of 2009. In 2009, 5 million mus of pilots were arranged, with the subsidy of RMB 100 per mu. In 2010, the scope of the forest tending subsidy was expanded and the pilot task was 20 million mus.

The SFGA and all provinces, autonomous regions, and municipalities have established a leading group on forest tending management, which clarified departmental responsibilities. Target responsibility was arranged at various levels, target management and performance appraisal were strengthened, thus forming a new pattern of joint efforts to promote forest management. The pilot management of forest tending subsidies has been comprehensively strengthened. The quality control system featuring self-examination at county level, verification, and acceptance at provincial level, and spot checks at national level has been established, while fund disbursement, survey design, public disclosure, technical training, contract management, supervision and guidance, inspection and acceptance, efficacy monitoring, and performance and penalty regime have also been put in place. With the focus on improving forest quality, and increasing the employment and income of forest farmers, measures have been taken to strengthen forest tending operations. Beijing, Zhejiang, Fujian, Hubei, and other provinces and cities have successively issued policies to increase local financial support for forest tending and low-function forest transformation, and expand funding channels for forest management; Liaoning, Hebei, Zhejiang, and others have revised forest management technical standards. Jilin, Guizhou, and others actively explore practical tending technical measures, with a variety of tending technology portfolios for expanding the effectiveness of forest tending management. In 2011, the area of young- and middle-aged forests under tending management was

7.3345 million hectares, an increase of 10.10% over 2010. The area of restructured low-function forest was 788,800 ha, an increase of 18.51% over 2010.

2.3 Forest Management Planning

In view of the low quality of China's forest resources, the National Outline on Forestry Protection and Utilization and the National A/R and Greening Plan, which were issued by the SFGA, make it one of the main tasks to carry out scientific management and implement forest quality engineering. Specifically, the focus is on the northeastern state-owned forest and the southern collectively owned forest, the goal is to greatly enhance the productivity of forestland and increase forest stock. Measures such as expanding the scale of forestland and rationally determining the incubation period were used to increase forest resources, enrich the supply of wood and forest products, and improve the carbon function of forest. For forestland in ecologically fragile areas, the cultivation of mixed and multi-level stratified forests is adopted to enrich biodiversity and enhance the stability of ecosystems. For forestland in important ecological areas, the focus is on cultivating large-diameter and long-period forest resources; for forestland in areas rich in hydrothermal resources, the focus is on intensive management and base management, with highlight on developing quality, precious and high-priced tree breeding bases to generate fast-growing, high-yield, high-quality, and efficient forest resources, and alleviate the structural problem of timber supply and forest products. By 2020, the national forestland productivity will reach 90 m^3/ha, while that of existing arbor forests will strive to reach 102 m^3/ha; the national forest stock will increase to more than 15 billion cubic meters, an increase of about 2.3 billion cubic meters compared with 2005. An increase of about 1.2 billion cubic meters compared with 2010; through forest management, consumption control, and other measures, the national forest reserve will strive to reach 15.8 billion cubic meters (SFGA 2010a, b).

By 2020, 7 million hectares of A/R under the canopy will be planned and 75 million hectares will be cultivated for forest tending (including low-yield forest restructuring). Among these, during the "12th Five-Year Plan" period, 3.5 million hectares of A/R under the canopy and 35 million hectares of forest tending management (including low-function forest restructuring) will be completed, respectively (SFGA 2010b). The so-called A/R under the canopy refers to A/R efforts for the stands with the following condition: pre-cutting renewal is needed and the canopy density is below 0.5 (excl.); unreasonable forest stands structure; the condition of natural regeneration is not enough; or the stand that can achieve normal growth if only the tree species be cultivated under the conditions of canopy shading (SFGA 2010b).

3 Scenario Analysis

3.1 Analysis Object

This scenario analysis is based on transforming low-function forest and uses forest tending as a way to restructuring low-function forests. The main approach to transforming low-function forest includes replanting, fencing, replacement, tending, adjustment, rejuvenation, and comprehensive transformation (SFGA 2007).

1. **Replanting**

Replanting is mainly for forest stand with canopy density of less than 0.5, plus unreasonable forest stands structure, and no natural regeneration conditions, or the stand that can only grow normally under canopy shading condition, such as residual forests, inferior forests, and low canopy density plantation. According to the current distribution of tree species, it can be divided into even replanting (forestland with relatively balanced forest distribution), plot replanting (existing forest trees are clustered, forest empty land and numerous forest gaps), and under canopy replanting (shade-resistant tree species). The planting density depends on the business purpose, the number of existing plants, and the reasonable density of the age group in which the forest is located. The density after replanting should reach 85% and above of the reasonable density of such forests.

2. **Fencing**

The fencing is mainly for stand with canopy density below 0.5, plus natural regeneration of young trees and seedlings, or broad-leaved trees distribution with natural regeneration ability, such as residual forests and inferior forests. Through fencing, external disturbances are reduced, and if necessary it is supplemented by manual measures to promote natural regeneration. The fencing period is 3–5 years in the south and 4–7 years in the north (Standardization Administration of the P.R.C. 2004).

3. **Replacement**

The replacement measures are mainly for residual forests, inferior forests, unsuitable forest species, pest-hazard forest, snow- and ice-stricken forest, degraded overripe forest, and improperly managed forest. All trees are cleared or gradually cut down by strips and blocks for timely regeneration.

4. **Tending**

Tending is mainly for inefficient pure forests, improperly managed forests, pest-hazard forests, and snow- and ice-stricken forests. For forests that need to adjust the composition, density or structure, the density is thinned, the quality species are kept, and light-penetrating cutting can be used; for stand that needs to adjust growth space, expand per plant nutrient extent and promote forest growth, growing cutting or selective cutting can be adopted; for pest-ridden forests, snow- and ice-stricken forests, damaged wood or pathogenic wood has to be completely removed, and stand sanitation has to be improved, which will contribute to restoring the inefficient forests to healthy development. Hygienic tending or selective cutting can be used.

5. **Adjustment**

The adjustment is mainly for inefficient forests or unsuitable species. The tree stand can be adjusted by means of thinning coniferous forest and supplementing with broadleaf trees, inserting coniferous species and tending broadleaf trees, and cutting coniferous species to preserve broadleaf trees. The one-time thinning intensity should not exceed 25% of the forest stand stock.

6. **Rejuvenation**

Rejuvenation is mainly for young- and middle-aged forests with poor growth. The methods of fertilization, sputum promotion, and soil loosening are used to restore the normal growth of the stand.

3.2 *The Analysis Method*

In view of the measures to restructure various low-function forests with the above-mentioned replanting, fencing, replacement, tending, adjustment, and rejuvenation, artificial low-function *Cupressus funebris* in the hills of central Sichuan and low-function *Cunninghamia lanceolata* and residual forests in southeastern Yunnan were taken as examples. Efforts were made to analyze the changes in biomass carbon reserve in the baseline and transformed scenarios. The baseline scenario here is the change in forest stand in the coming years with no restructuring measures taken. Whether it is baseline scenario or transformed scenario, the land use pattern is forest-land, thus according to IPCC Tier 1, the change of soil organic carbon can be assumed to be zero. Although some measures may result in changes in carbon reserve in dead wood and litter under baseline or project scenario, they are not included in this analysis.

In baseline scenario and project scenario, the estimation of forest biomass carbon stock changes is based on the biomass expansion factor method (BEF method), which converts the unit standing volume into biomass and carbon reserve.

$$C_t = V_t \cdot D \cdot BEF \cdot (1 + R) \cdot CF$$

where C_t, biomass carbon reserve per unit area, tons of carbon dioxide per hectare; V_t, volume of forest stock per unit area, cubic meters per hectare; D, average trunk density, tons of dry matter per cubic meter; BEF, biomass expansion factor (ratio of aboveground biomass to trunk biomass), dimensionless; R, forest biomass ratio, dimensionless; CF, forest carbon content (0.47 available), tons of carbon/ton of dry matter.

Once harvesting occurs, whether it is thinning or main cutting, it is assumed that the biomass carbon reserve in the harvested trees is immediately released into the atmosphere. Regardless of the fate of the harvested wood, the carbon reserve in the wood product is not calculated.

When analyzing the changes in biomass carbon reserve under project scenario and baseline scenario, the comparative analysis is based on long-term average carbon reserve. The so-called long-term average carbon reserve refers to the average carbon reserve level over a long period of time, such as 30 years.

If there is no main cutting during the crediting period, the long-term average carbon reserve will be the carbon reserve at the end of the crediting period.

If there is main cutting during the crediting period, but selective cutting (each time no more than 50% of the stock) is used, the long-term average carbon reserve is the average of that before and after the last selective cutting prior to the end of the crediting period.

If there is main cut during the crediting period and is all clear-cutting, the long-term average carbon reserve is 50% of that before the last main cutting prior to the end of the crediting period.

4 Scenario Analysis of Case Study

4.1 Restructuring of Low-Function and Over-Density Cupressus Funebris Forest in the Hills of Central Sichuan

1. **Project background**

Sichuan Province started the construction of the Yangtze River watershed Forest phase in 1989. By the end of 1996, it had built 1.733 million hectares of shelter forests, mainly distributed in the hilly agricultural areas of Sichuan, and the forest stand was mainly *Cupressus funebris* forests. Although the shelter forest has greatly improved the ecological environment of the hilly area in central Sichuan, and significantly increased the forest coverage rate, there were large area of low-function forest due to poor site conditions, excessive A/R density, unreasonable tree species, poor seedling quality, and low management level.

Cupressus funebris is an important local conifer species in central Sichuan and has a long history of cultivation in the area. However, because the A/R investment was low at the time, and cypress was the cheapest seedling, it naturally became the main species for developing the shelter forests in the region, thus forming a large area of singular cypress forest.

Due to the fact that in the lower hilly areas of central Sichuan and northern Sichuan, the purple shale rock gives rise to coarse brown purple soil, and the soil texture is coarse, with poor fertility and weak water retention capacity. The bedrock features purple sandy shale, which is permeable rock type and makes A/R extremely difficult. In order to ensure the survival rate and achieve the goal of greening as soon as possible, local governments promote the A/R technology of "multi-density plantation", namely, planting 4–6 cypress trees in each tree hole.

Due to the fact that people focus on A/R over management, there is no fund for tending the forest stand, and the tending management is either extensive or non-caring for long time after A/R, resulting in excessive forest stand density, deteriorating environmental conditions for growth, and poor growth of young forests. At present, about 70% of the forest has too much density.

Unreasonable shaving seriously affects the growth of trees. When local residents use the timber, they cut big trees instead of small ones, straight types instead of bend ones, and quality species instead of worse ones, which caused the de facto elimination. The remaining forest has poor ecological and economic benefits.

2. **Restructuring mode**

According to a survey conducted in Langzhong and Qu County, the age of man-made low-function *Cupressus funebris* forest was 13–38 years and the density was 3000–6600 plants/ha. Taking the typical plot forest as an example (25 years old, average DBH 8.50 cm, average height 7.3 m, stock 74.69 m^3/ha), the following three restructuring scenarios will be adopted:

Mode: even thinning.

Even thinning is mainly carried out in forestland where the site conditions are poor, the slope is steep, and the soil is relatively barren. It follows the principle of timeliness and appropriateness, chopping inferior to keep quality species, slashing dense types to retain sparseness, cutting smaller trees to maintain large ones, and keeping balance and evenness. Within 50 years of the project's crediting period, two evenness thinning were planned, respectively, in the 1st year and the 11th year. For the first time, the density of the existing forest stand was reduced to 2,700 plants/ha by thinning and the second time was down to 2,100 plants/ha.

Mode: strip thinning and adjustment.

Strip thinning and adjustment are mainly carried out in forests with better site conditions, and are operated in a manner of cutting 3 m after every 5 m of interval along the contour line. Even thinning is done to the 5 m interval, with the same intensity as the even thinning mode. A line of local broad-leaved tree species is planted in the clear-cutting zone, for example, Alnus cremastogyne, camphor or Robinia pseudoacacia, with a spacing of 2 m (1667 plants/ha). The planted broad-leaved trees will be harvested in 10 years, with intensity of 20%; the main cutting age is 20 years. During the crediting period, no cutting will be done to Cupressus funebris.

Mode: species change.

Similar to the mode of adjustment measures, the existing low-function forests are cut off and replaced by local broad-leaved species such as Alnus cremastogyne, camphor or Robinia pseudoacacia, with spacing of 2 m (1667 plants/ha). The broad-leaved species are harvested in 10 years, with intensity of 20%, and the age of main cutting is 20 years.

3. Carbon potential

Using the survey data of forest resources in Sichuan Province, the breast diameter growth process of different density grades was calculated, the average single plant volume of the scenario and the baseline scenario was measured by the DBH calculation, and the number of plant per unit area and the accumulated growth amount were calculated according to the distribution of the number and diameters grade. Finally, the carbon reserve per unit area and long-term average carbon reserve during the crediting period were calculated. The growth stock of the adjusted or replaced broad-leaved tree is calculated by using the volume growth equation, followed by carbon reserve calculation.

The results show that although the tree structure is improved by reducing coniferous species for more broad leaves, and the wood yield is also improved, the long-term carbon reserve is much smaller than that of even thinning, since the rotational cutting period of broadleaf trees is shorter. After the tree species are replaced by broadleaf species, their long-term carbon reserve is smaller. Long-term carbon reserve for baseline scenarios, even thinning, strip thinning, adjustment, and species replacement were 70, 100, 68, and 33 tons of carbon per hectare, respectively. Only even thinning has a net carbon benefit, while the adjustment and replacement modes have no net carbon benefits to mention (Fig. 5).

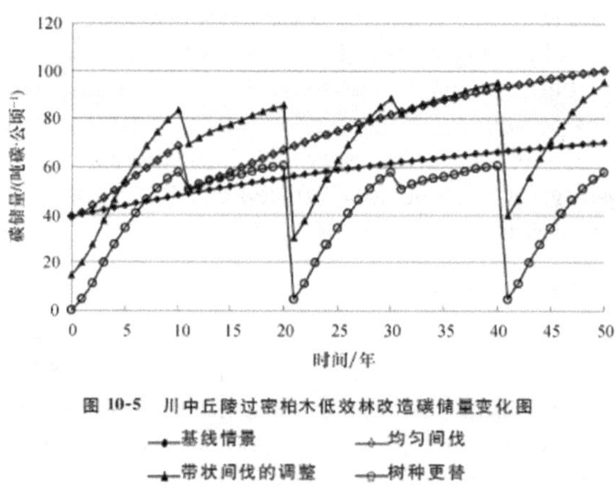

碳储量/（吨碳·公顷$^{-1}$）	Carbon reserve/(tons of carbon·hectare^{-1})
时间/年	Time/year
基线情景	Baseline scenario
均匀间伐	Even thinning
带状间伐的调整	Strip thinning adjustment
树种更替	Species replacement

Fig. 5 Figure of carbon stock change for restructuring low-function and over-density cypress in the hilly areas of central Sichuan

Many studies have shown that thinning does not increase the carbon reserve of stands. However, a large number of surveys have shown that since the thinning increases the light penetration of the stand, it promotes the development of understory vegetation, and the biomass of understory vegetation can be increased to 5–10% of the total (Sheng and Zhang 2011). However, other studies have shown that thinning can contribute to forest growth and arbor carbon reserve (Zhang 2009; Luo et al. 2009). Therefore, in the long run, whether thinning can be conducive to forest carbon reserve depends on environmental conditions and stand factors such as site conditions, thinning intensity, stand age, and crediting period. The hilly area in central Sichuan suffers serious soil erosion and loss, and most of the local *Cupressus funebris* forests are for public welfare purposes. Therefore, regardless of environmental impacts or carbon benefits, it is not advisable to adopt species replacement after low-function forest clear-cutting.

For strip adjustment and species replacement, if the adjusted tree is long-period species, or for cultivating large-diameter precious materials and with rotational period longer than the crediting period, a certain amount of net carbon will be generated.

4.2 Restructuring Low-Function Forest in Southeast Guizhou

1. **Project background**

Southeast Guizhou is a key collective forest area in southern China and is one of the 28 major forest zones in the country. The natural conditions are suitable for tree growth. There are 1.822 million hectares of forestland in Guizhou, and the forest coverage is 62.78%, which is close to the theoretical maximum forest coverage. However, the forests quality in southeast Guizhou is relatively poor. The average standing volume per hectare of arbor species in commercial forests is only 78.15 m^3, which is lower than the national average of 86 m^3. The man-made forest is mainly coniferous *Cunninghamia Lanceolata* and *Pinus massoniana*, accounting for 82% of the total. The stand stability and ecological benefits are poor, and the ability to withstand natural disasters is weak. Currently, there are about 200,000 ha of inefficient forests in the Province. The main reasons for the formation of inefficient forests are as follows.

As an important timber supply base of the country, southeast Guizhou provides a large amount of timber for the country, but the long-term lack of investment, the insufficient infrastructure, and inadequate forest management have made it difficult to improve the quality of forests.

In the early 1990s, the Guizhou Provincial Party Committee and the provincial government made a strategic decision to "eliminate the barren hills in seven years and green the Province in ten years." The southeast regional government also worked out a greening strategy of "eliminate the barren hills in seven years and greening the region in ten years." Due to the heavy task, the investment per unit area is low. The

A/R cost is RMB 40–80 per mu, and social investment is only a few RMB in seedling costs. The greening has become the top priority of forest management, but the quality of forest management failed to reach the expected level of improvement. At present, the plantations cultivated at this stage have all entered the middle-aged forest stage. Due to the serious shortage of funding, the financial and technical inputs in seedlings, soil preparation, and tending are not sufficient enough, the forest production process cannot be guaranteed, resulting in low forest quality. Most of the forests planted during this period have become the target of restructuring.

The traditional approach of "taking down big trees": in southeast Guizhou, there is way of forest utilization called "taking down big trees". People regard forest as their own bank. In addition to large-scale harvesting for house construction, there are other situations in which people will harvest large-diameter trees according to their needs and sell them in exchange for cash, such as weddings, funerals, and school education. After such harvesting, the rest are only small-diameter trees and residual trees. The more such practice is used, the lower the unit area stock will become.

Inadequate tending of young forests and serious dereliction of tending and thinning. Sometimes first thinning was conducted in 13–14.

According to the abovementioned features in southeast Guizhou, the Regional Party Committee and the Government decided to improve forest quality as a starting point, and determined to carry out low-function forest restructuring. They make the restructuring as one of the key measures to increase farmers' income, and make use of the forest restructuring as an opportunity to increase the output per unit area of forest land, so that local people can rely on mountains for food and security. To this end, the regional forestry bureau commissioned the Central Forestry Investigation and Design Institute of the SFGA to map out the overall plan for low-function forest restructuring. During the period of 2011–2015, more than 200,000 ha of low-function forests will be restructured, of which 137,000 ha will be replaced (clear-cutting) and 56,000 ha will be rehabilitated (including 39,400 ha of thinning and replanting and 164,000 ha of A/R under the canopy), 8,100 ha (including 1,331 ha of weeding, 4,946 ha of fertilization, and 1,858 ha of enclosure for A/R) will be rejuvenated. In terms of forest species to be restructured: 112,000 ha for short-rotation industrial raw material timber, 34,000 ha for fast-growing high-yield timber forest, 34,000 ha for economic forest, 16,000 ha for general public welfare forest, and 4,000 ha for bamboo forest.

According to the survey on the status of forest resources management in 16 counties (cities) in southeast Guizhou, based on the *Technical Regulations on Restruction of Low-function Forest* (LY/T1690–2007) and the *Technical Regulations on Improvement of Low Yield Timber Forest* (LY/T1560–1999) issued by the SFGA, also in combination with the actual situation of the region, the planned restructuring criteria are the forest stands above the middle age and with a canopy density less than 0.3; the middle-aged forest with average annual growth less than 0.3 m^3 per mu; low-quality and low-yielding economic forest with an average annual output of 70% lower than the local average in the past 3 years; low-function shelter forest with forest canopy density less than 0.3.

2. **Restructuring mode**

(1) Species change

This mode accounts for 68.2% of the total planned area. It will transform the existing low-function forest into short-rotation industrial raw material forest, fast-growing and high-yield forest, economic forest, and bamboo forest. The main species are *Chinese fir*, *Masson pine*, *cypress*, *eucalyptus*, *Liquidambar formosana*, *Liriodendron chinense*, and camphor; economic forests are mainly oil tea and tea leaves.

(2) Thinning adjustment

This mode accounts for about 19.6% of the total planned area. It is aimed at low-function forests for general public welfare purposes, strips, or blocks thinning except for suitable trees or damaged trees. Tree species that are adapted to climate conditions and soil conditions for planting are introduced. The restructuring intensity is controlled within 20% of the stock and 50% of the number of plants.

(3) Replanting

This mode accounts for about 8.1% of the total planned area. For the sparse forests where the forests for general public welfare purposes are fragmented, even replanting or block replanting is adopted according to the size and distribution of stand gaps. In general, 1,000–2,000 trees/ha are replanted in natural forests, and 1,500–2,000 tree/ha are replanted in plantation. After the restructure, a mixed planted and natural forest is formed.

(4) Fencing

For the stand of public welfare purposes featuring mixed bush and broad-leaved trees in harsh environment, steep slopes and difficult to recover conditions, for sparse forests featuring *Masson pine* at steep slopes and with 10–15 trees per mu, and for low-yield and Low-function forests mainly caused by human disturbance, hills enclosure is used for A/R, anthropogenic activities are strictly controlled and management and protection are strengthened. The results after enclosure are stratified mixed forests.

3. **Carbon benefit analysis**

(1) Species replacement

According to the overall plan, the short-rotation industrial raw material forests of *Masson pine*, *Chinese fir*, and broad-leaved trees can amount up to 143–179 m^3/ha in the 10th year of harvesting, and the stock of fast-growing and high-yield timber forests can reach 267,271 m^3/ha in the 16th year of harvesting. Suppose the proposed forest is divided into 15-year-old low-function forest of *Masson pine* and *Chinese fir*, with a volume of 60 m^3/ha. The average annual growth is 3.75 m^3/ha (0.25 m^3 per mu per year). From this, the carbon stock changes in the restructuring scenario and the baseline scenario during the 30-year crediting period can be derived. The results show that, compared with the long-term carbon stock (56.3 tons of carbon/ha), the carbon stock for the restructured short-rotation industrial raw material forests and fast-growing high-yield timber forests (25.8 tons of carbon/ha. and 37.7 tons of carbon/ha, respectively) are much lower, and there is no net biomass carbon in the restructure scenario (Fig. 6). This is mainly related to the main harvesting in the restructure scenario. If the carbon reserve in wood products is added, even if there is a net carbon sink, it is very limited.

However, if the restructured stands are cultivated for long-period and large-sized timbers, and with no significant clear-cutting during the 30-year crediting period, then biomass carbon stock (Fig. 7) will be significant. Compared to the baseline scenario, the total carbon per hectare during 30 years can reach 60 tons of carbon per hectare.

If the selective cutting method is adopted, it is assumed that the selective cutting intensity is 50%, that is, 50% of the forest rotation period is doubled (short-rotation industrial raw material forest and fast-growing high-yield forest are 20 years and 32 years, respectively), and then short-rotation industry, the raw forest model (long-term average carbon stock of 52 tons of carbon/ha) still has no net carbon sink, while the fast-growing and high-yield forest model (long-term average carbon stock of 70.2 tons of carbon/ha), compared with the baseline scenario, the 30-year carbon pool about 14 tons of carbon per hectare (Fig. 8).

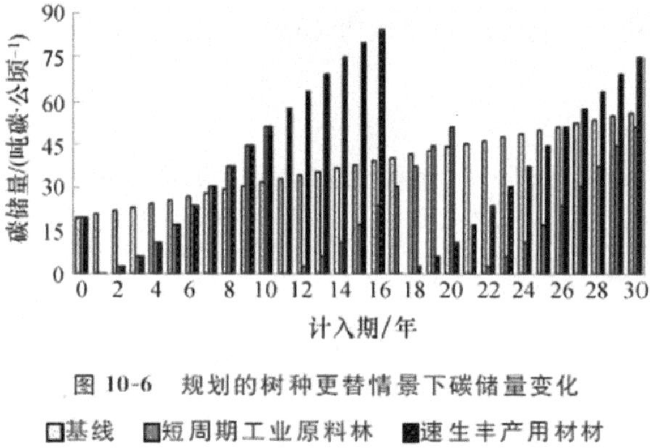

碳储量/（吨碳·公顷$^{-1}$）	Carbonstock/(tons of carbon·hectare^{-1})
计入期/年	Crediting period/year
基线	Baseline
短周期工业原料林	Short-rotation industrial material forest
速生丰产用材林	Fast-growing high-yield forest

Fig. 6 Change in carbon stock in the planned species replacement scenario

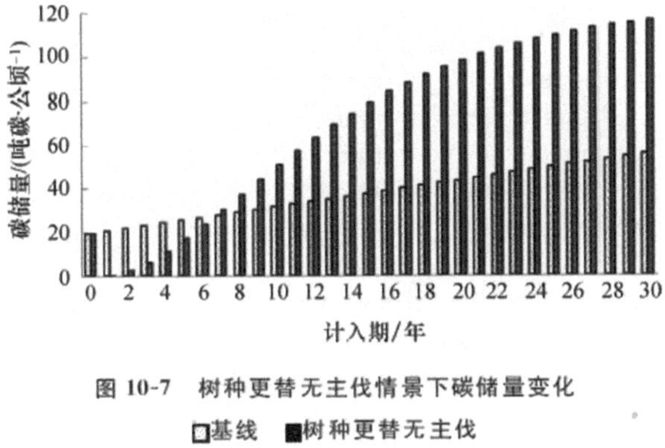

碳储量/（吨碳·公顷$^{-1}$）	Carbonstock/(tons of carbon·hectare^{-1})
计入期/年	Crediting period/year
基线	Baseline
树种更替无主伐	Species replacement, no clear cutting

Fig. 7 Carbon stock change in species replacement and no clear-cutting

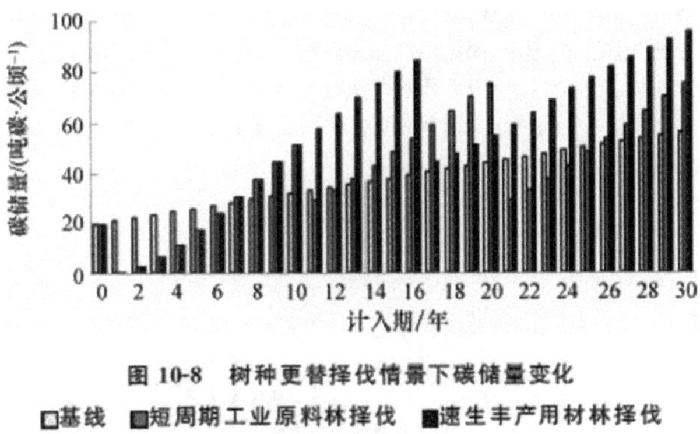

碳储量/（吨碳·公顷$^{-1}$）	Carbon stock/(tons of carbon·hectare^{-1})
计入期/年	Crediting period/year
基线	Baseline
短周期工业原料林择伐	Selective cutting for short-rotation industrial raw material
速生丰产用材林择伐	Selective cutting for fast-growing high-yield timber

Fig. 8 Carbon stock in selective cutting scenario of species replacement

(2) Thinning adjustment

Since this mode target restructuring of low-function forest for public welfare purposes, it is assumed that there is no main cutting in 30 years. Suppose that each thinning is 20% of the stock, and cutting and replanting is done once in every 5 years, the baseline scenario is still assumed to be 15-year-old *Chinese fir* or *Masson pine* low-yield forest with a volume of 60 m^3/ha and an average annual growth of 3.75 m^3/ha (0.25 33 m^3 per mu each year). From this, the carbon stock changes in restructuring scenario and baseline scenario during the 30-year crediting period can be estimated. In the first 10 years, the carbon stock in restructuring scenario is lower than that of baseline scenario. From the 13th year onward, there is net carbon in restructuring scenario than that of the baseline scenario. During the 30-year crediting period, the cumulative carbon can reach 35.1 tons of carbon per hectare (Fig. 9). However, during the crediting period, if the replanted forests are managed as short-rotation industrial raw material forests or fast-growing high-yield forests, there will be no net carbon sinks in the restructuring scenario.

(3) Replanting and enclosure

Since replanting and enclosure measures are mainly for low-function forests of public welfare purposes, generally there will not be harvesting during the 30-year crediting period. Therefore, compared with the baseline scenario, net carbon sinks will be generated. For replanting, the amount of carbon depends on the density of existing low-function forests. The smaller the density, the larger the amount of replanting will be, and the greater the net carbon will be. However, if main cutting is carried

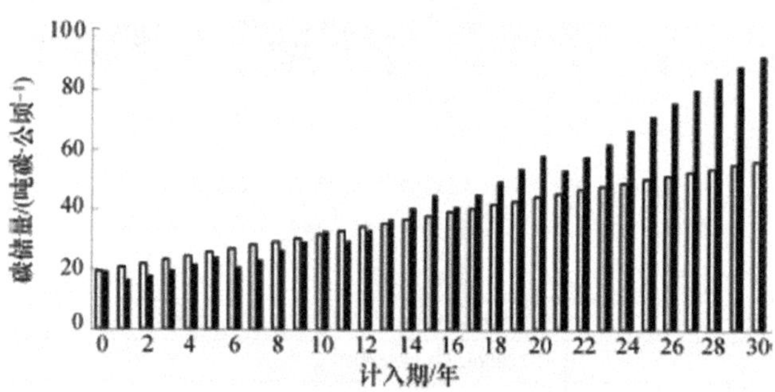

图 10-9 间伐调整情景下碳储量变化

□基线 ■间伐调整

碳储量/（吨碳·公顷-1）	Carbon stock /(tons of carbon·hectare^{-1})
计入期/年	Crediting period/year
基线	Baseline
间伐调整	Thinning adjustment

Fig. 9 Carbon stock change in thinning adjustment scenario

out during the crediting period, the net carbon will be very limited or even no net carbon sink. For enclosure measures, the amount of carbon sinks depends on factors such as site conditions, provenance, and natural regeneration capacity.

References

Deng D, Zhang X, Yan W, Mu C (2010) Summary of research on restructuring low-efficient forest 23 (4):65–69

Luo Z, Hou B, Xiang C, Chen J, Luo X, Xie D, Changlong M (2009) Restructuring low-efficient Shelter Forest of *Cupressus funebris* in the Basin, low mountain and hilly region of Sichuan. J Cent South Univ For Technol 29(6):82–87

Peter-Stanley M, Hamilton K, Marcello Thomas, Sjardin M (2011) Back to the future: state of the voluntary carbon markets 201. In: A report by Ecosystem Marketplace & Bloomberg New Energy Finance, 2 June

SFGA (2007) Technical regulations on restruction of low-function forest. Standard on Forestry of P.R. China: LY/T1690–2007

SFGA (2009) Report on China's forestry resource—7th national forestry resource inventory. China Forestry Press, Beijing

SFGA (2010a) Outline of the protection and utilization planning of forest land (2010–2020)

SFGA (2010b) Outline of national A/R and greening plan (2010–2020)

Sheng W, Zhang X (2011) Forestry cultivation and carbon sink. In: Liu Y, Wang H, Zhang X (eds) Climate change and forestry carbon sink. China Meteorological Press, Beijing

Standardization Administration of P.R.C. (2004) Technical regulation on forest conservation: GB/T15163–2004

Zhang Z (2009) Research on cost-benefit of carbon sequestration in A/R and reforestation in Guangxi. Chinese Academy of Forestry

Zhang X (2011) Contribution of forest management credits in Kyoto Protocol compliance and future perspectives. Adv Climate Change Res 2(4):171–177

Zhang X, Shuhong W (2010) Theory and practice of forestry carbon project. China Forestry Press, Beijing

Obstacles, Experiences, and Recommendations on Forest Carbon Projects

Chunfeng Wang and Caifu Tang

Summary
This chapter is focused on the CDM A/R project developed and implemented in China. It analyzes the policy barriers, capacity barriers, investment and financial obstacles, market, and risk barricades facing forest carbon projects in China, and summarizes the experience gained. Finally, relevant international and domestic countermeasures and recommendations are proposed from policy and technical perspectives.

In terms of policy barriers, the obstacles faced by forest carbon projects include: narrow scope, the first commitment period is limited to A/R, the land eligibility requirement is much too strict, thus the potential is slight; using tCER and lCER to solve non-permanence issues, causing low forest carbon price and restricted buyers' market, thus the development of forest carbon trading is also limited; the project boundary requirements are very rigorous, the cost of boundary identification is high and risky, and the coordinate system used is difficult to be connected with that of international common system; China promotes large-scale A/R and forest management, which will face greater challenges when demonstrating the additionality of policies and the common practice of projects; the rules, procedures, and methodologies of forest carbon projects are complex, difficult in development, high in cost and risk, long in development cycle; the land user right is unclear, and with frequent disputes, which brings great challenges to the development and implementation of forest carbon project; the wood product carbon pool is not included in the methodologies, so that the forest carbon under sustainable management is greatly compromised; China requires that the project implementing entity shall be a Chinese-funded or

C. Wang (✉)
State Forestry and Grassland Administration, Beijing, China
e-mail: chunfengw@126.com

C. Tang
Sichuan Green Carbon Ltd., Chengdu, China
e-mail: 631592345@qq.com

Z. Lu et al. (eds.), *Forest Carbon Practices and Low Carbon Development in China*, https://doi.org/10.1007/978-981-13-7364-0_9

Chinese-holding company, which is not conducive to the development and implementation of forest carbon projects. In terms of technical capacity barriers, the problems encountered are as follows: few domestic experts in developing methodologies for forest carbon projects; the implementing bodies, local authorities, and technical personnel have insufficient understanding of the rules and requirements of forest carbon projects; project implementation bodies and relevant professional and technical personnel are negligent in tracking the update of carbon standards, guidelines, instructions, methodologies, and their tools; lack of awareness and ability to gather, collect and archive relevant evidence; lack of relevant carbon measurement equations and parameters; insufficient experience in A/R on difficult sites; and so on. In terms of investment and financial obstacles, forest carbon projects face high investment, low efficiency, slow effect, and high transaction costs; it is difficult to use carbon income for solving the investment barriers faced by A/R projects. In terms of risk, forest carbon projects face high fire risks, extreme weather events (such as rain, snow, severe coldness, and extreme drought) and risks of land use change. The experience of successful forest carbon projects over the past few years has shown that some good practices can help reduce the barriers and risks in the development and implementation of forest carbon projects. They include: maintaining close communication with relevant national authorities; establishing efficient operational management structure to give full play to the role of local forestry authorities; setting up a composite expert group composed of international, national, and local experts who are familiar with the project area and various subjects, so that they can learn from each other's strengths and work together; creating a reasonable business and community cooperation model, increasing community participation and benefits; developing systematic training and publicity campaign for different audiences, improving project management and technical capabilities, and raising awareness to forest carbon; using participatory design and monitoring, and communicating and coordinating closely with community protection; developing and implementing strict quality assurance and quality control procedures; gathering, collecting and archiving evidence at any time, etc. The final recommendations include: expanding the scope of forest carbon projects at international level; relaxing land eligibility requirements for CDM A/R projects; adopting new non-permanence solutions; giving project boundaries more flexibility; simplifying project rules, approaches, and procedures; including wood product carbon pools in the measurement of forest carbon. At domestic level, regard forest carbon projects as a priority area for international and domestic carbon trading; fully absorbing the experience of CDM A/R projects and improving on the deficiencies; identifying with international and domestic voluntary carbon standards (such as VCS, Carbon Fix-Standard, panda standards, etc.) regarding methodology and methodological tools for forest carbon, approving simplification and revision if necessary; encouraging and providing necessary financial support to develop forest carbon methodology suitable for China; develop additionality criteria in line with China's national conditions, social conditions, and forest conditions; encouraging the adoption of multiple benefit standards to enhance social and environmental benefits of the project; further promoting the reform of collective forest ownership system to reduce the risk of land ownership disputes; allowing local forestry authorities to directly declare and implement international forest carbon projects. At the same time, vigorously carrying out publicity and technical training on forest carbon projects, conducting relevant basic research, and setting up forest resource data sharing mechanism.

1 Obstacles Analysis

1.1 Policy Obstacles

1. Project area

There are many types of forestry activities that generate net carbon sink, such as A/R, forest management, wetland restoration, vegetation restoration, deforestation, and forest degradation. However, only A/R is qualified CDM project activity. Although the volume of forest management, wetland restoration, deforestation, and forest degradation projects in the voluntary carbon market increased rapidly in recent years, it is much smaller than the CDM market, and there is no successful case of wetland recovery project in China. In particular, there are not much forests suitable for A/R in China, and the existing forests are of poor quality, the wetlands are seriously degraded, and the natural conditions are poor. Therefore, the forest management, vegetation restoration, and wetland restoration carbon projects enjoy relatively greater potential to be tapped with.

2. Land eligibility

According to the Resolution 16 on UNFCCC, land use change and forestry adopted at the 1st COP of the Kyoto Protocol, only forest-free land since December 31, 1989, is qualified CDM A/R project activity (UNFCCC 2005a, b). Since the founding of the People's Republic of China, especially since the Reform and Opening-up, the Chinese government has vigorously promoted A/R, greening and the elimination of barren hills. The forest area has increased substantially, and the area suitable for forests has been rapidly reduced. According to the 7th National Forest Resources Inventory from 2004 to 2008, the national suitable forest area is about 44 million hectares (SFGA 2009), of which only a small part may meet the requirements for land eligibility for CDM A/R projects, because: About 60% of these suitable land is distributed in Inner Mongolia and five northwestern provinces, accounting for 35.9% and 23.8%, respectively (SFGA 2009). Although most of these are forest-free land since 1989, due to the worsening natural conditions, many suitable forestlands are no longer suitable for growing arbor forest, but more suitable for shrub forest. Creating shrubs is not qualified CDM A/R project activity. In voluntary carbon standards for A/R, the Panda Standard and VCS regulations state that the land shall be forest free

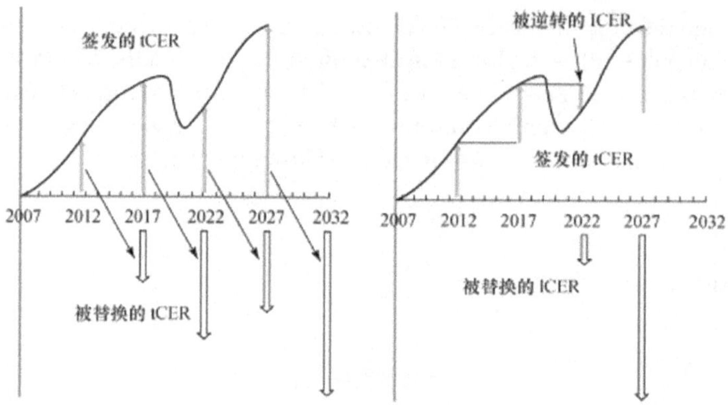

签发的 tCER	Issued tCER
被替换的 tCER	Replaced tCER
被逆转的 ICER	Reversed ICER
被替换的 ICER	Replaced ICER

Fig. 1 Chart of tCER and ICER (Zhang and Shuhong 2010)

for at least 10 years prior to the start of the project; the Carbon Fix Standard stipulates that the lands for A/R project must use forest free since January 1, 1990, or no forest at least 10 years before project commences (Zhang et al. 2009).

3. Non-permanence

The so-called non-permanence means that the carbon dioxide absorbed by forests created during A/R activities will be released into the atmosphere due to anthropogenic or natural reasons such as logging, fire, pests and diseases and deforestation, reversing the greenhouse gas benefits of CDM A/R project activities. In fact, all forest carbon projects, whether A/R, forest management, vegetation restoration, or reduced deforestation, and forest degradation have unsustainable issues. To address the problem, the CDM A/R project uses tCER and long-term certified emission reductions (lCERs).

Since both tCER and lCER need to be replaced, this seriously affects the CER price and the enthusiasm of purchasers. The price of CERs generated by CDM A/R projects is much lower than those generated by CDM emission reduction projects in industry and energy sectors. Comparatively speaking, China's CDM-based A/R projects have higher transaction prices, but still lower than CER prices in the field of emission reduction.

Definitions and interpretations of tCER and CER

tCER refers to the CER that remains valid at the end of the next commitment period upon signature of commitment period (UNFCCC 2005a, b). In other words, with respect to the tCER issued for the first commitment period (2008–2012), if the second commitment period is still 5 years (2013–2017), the tCER must be replaced before the end of 2017 by using the assigned amount unit (AAU), emission reduction unit (ERU), CER, remittance unit (RMU) or other tCER generated by similar projects; if the second commitment period is 8 years (2013–2020), the tCER must be replaced before the end of 2020. The replaced tCER can continue to be used for future commitment period, that is, the project owner can re-issue and sell the replaced tCER after re-verification and certification.

lCER refers to the CER that remains valid at the end of the crediting period (UNFCCC 2005a, b). In other words, the lCER issued during any commitment period is allowed to be replaced with the lCER generated by AAU, ERU, CER, RMU or other similar projects before the end of the crediting period. As long as the net anthropogenic greenhouse gas removals by sinks generated by CDM A/R project activities are not reversed and project participants can submit monitoring reports to the CDM-EB on time, the validity period of lCER can be maintained until the end of the crediting period. The main differences between tCER and lCER are reflected in the following aspects (Fig. 1) (Zhang and Shuhong 2010):

(1) Different number of issuance

The amount of tCER issued after each verification and certification is the cumulative net anthropogenic GHG removals by sinks by CDM A/R project activity since the start of project activity. The amount of lCER issued after each verification and certification is the net anthropogenic GHG removals by sinks by CDM A/R project activity since the previous verification and certification. If the net anthropogenic GHG removals by sinks has a negative value (net discharge), then the lCER issued in the past has been reversed.

(2) Different validity periods

The tCER issued after verification and certification will expire at the end of the next commitment period for which it is issued. For example, for the tCER issued for the first commitment period of the Kyoto Protocol (2008–2012), if the second commitment period is still 5 years (2013–2017), the tCER will expire at the end of 2017, namely the validity period is 5 year; if the second commitment period is 8 years (2013–2020), the tCER will expire before the end of 2020, that is, the validity period is 8 years. Since the lCER issued after verification and certification will become invalid at the end of the crediting period, the validity period of lCER will vary depending on the time of verification and certification. As long as the net anthropogenic GHG removals by sinks generated by CDM A/R project activities are not reversed and project participants can submit monitoring reports to the CDM-EB on time and keep on verification and certification, the validity period of the lCER can be maintained until the end of the crediting period. In the case of an updatable crediting period, lCER is valid until the end of the last crediting period. For example, for a fixed crediting period of 20 years, if the first verification and certification is in the fifth year after the start of the project, the lCER issued after the first verification and certification is valid for 15 years. The validity period of the lCER issued after the second verification and certification is 10 years, provided that the issued lCER has not been reversed. In the case of a crediting period that can be updated twice in 20 years, the first and second issued lCERs can be valid for 55 and 50 years respectively. If a reversal occurs, the reversal portion of the lCER is valid for only five years.

(3) Different replacement conditions

Whether it is tCER or lCER, they must be replaced. However, tCER is only replaced when it fails. If the replaced tCER is not reversed, it is re-issued after verification and certification, and lCER has to meet the following three replacement conditions:

Invalidation: The lCER issued for any commitment period must be replaced before the end of the crediting period. If it is an updateable crediting period, it can be replaced at the end of the last crediting period.

Reversal: If the lCER issued for a certain commitment period is reversed after issuance, namely, the next verification and certification will indicate that since the last verification, the net anthropogenic GHG removals by sinks has been reversed, the reversed part of lCER has to be replaced immediately.

Failure to submit certification report on time: no verification report was received within 5 years after the last certification, and no certification report received within 120 days after the project participant received the notice from the Executive Board. All lCERs issued have to be replaced.

(4) Different values

Since tCER and lCER have different expiration dates, the value of tCER is lower than that of lCER.

(5) Different costs

The cost of lCER and tCER varies greatly depending on the number of issuances. Taking the registered CDM "the reforestation project at the Pearl River Basin of Guangxi in China" as an example, it is assumed that the future commitment period is 5 years. The cost of carbon sequestration in the respective case of lCER and tCER indicates that, except for the first commitment period where lCER and tCER have the same cost, in all future commitment periods, the cost of lCER is significantly higher than that of tCER, and the cost of lCER is showing increasing trend, while the cost of tCER is decreasing (Zhang et al. 2009).

4. Project boundary

Forest carbon projects are required to identify detailed project boundaries and provide GIS-based shapefiles for project plot boundaries. China's forest-suitable land is very limited and mostly located in remote areas. Especially in the south, many of such land are located in the inaccessible mountains. The project plots are scattered around steep mountains, with thick thorns and cliffs. This poses a huge challenge to the GPS boundary prospection due to the high cost and risk for operation. Although on-site boundary mapping can be used to avoid these problems, it may cause greater errors, such as incorporating adjacent forestland into the project boundary, causing serious land eligibility problems.

Another challenge facing China's project plots is the coordinates system. The coordinates system used in China is usually the Xi'an 80 and Beijing 54 types, or even local coordinates system, while the international common coordinates system is WGS84. These systems are interchangeable. However, since several parameters required for coordinate conversion are confidential, the relevant forestry survey and design agencies and project owners cannot obtain these conversion parameters, thus they cannot convert boundary coordinates into an international common coordinate system, which brings great difficulty to verification and verification. For example, "the reforestation project at the Pearl River Basin of Guangxi in China" uses the Beijing 54 coordinates system to define the project boundary and the coordinates of the center point of the monitoring plot. However, for the first verification, DOE used the WGS84 coordinates system, resulting in a situation where the measured project boundary and monitoring system coordinates are systematically offset from the monitoring coordinates, which brings great difficulties to verification.

5. Additionality

Both the CDM A/R project and forest carbon project under the voluntary carbon standard require that the net carbon sinks generated by projects be additional in relation to the baseline scenario. To this end, the project developer, when developing a project, must demonstrate and provide transparent and verifiable information to demonstrate the additionality of the project.

In fact, the additionality implies several aspects, namely, environmental additionality, capital additionality, investment additionality, technical additionality, and policy additionality, which require demonstration of project additionality from investment, capital, technology, and policy aspects. The CDM A/R project is required to demonstrate the additionality of project in accordance with the steps in "Comprehensive tool for Baseline Scenario Identification and Additionality Assessment of CDM A/R Project Activity,", including project start date, identification of land use alternatives, barrier analysis, investment analysis, and common practice analysis.

The Panda Standard requires a "triple test" (in compliance with legal and regulatory requirements, surpassing general practices, and facing investment, technical or institutional barriers) or performance standards (beyond legal and regulatory requirements and performance standards set in the Panda Standard). The comprehensive tool for baseline scenario identification and additionality assessment for CDM A/R project activities may also be used for A/R projects.

When demonstrating the additionality of China's forest carbon projects, the biggest challenge is policy additionality and common practice analysis. The so-called policy additionality means that forest carbon project activity cannot be mandated by the relevant laws or regulations or policies of the state or any department. Otherwise, such project activity belongs to the baseline scenario without any additionality, unless project participants can prove that relevant laws, regulations or policies are not universally implemented. These laws, regulations, or policies are usually embodied in the form of government plans, financial allocations, and general policy objectives. For CDM projects, the timeline for these laws, regulations, or policies is November 11, 2001, when the Marrakesh Accords adopt the CDM project approach and procedures. In demonstrating baseline scenario of a project, the relevant state and/or sectoral policies and regulations (CDM-EB 2008) that are drafted and implemented after the timeline and are conducive to A/R activities may not be taken into account. That is to say, the state or sectoral laws, regulations or policies that are beneficial to A/R activities, which were implemented on November 11, 2001, will not affect the additionality of CDM A/R project activities. In the past few decades, China has implemented large-scale A/R and forest protection programs, especially the six major forestry projects implemented in the past decade, all of which began before November 11, 2001. Therefore, how to distinguish project activities from national or local forestry projects and related investments and provide convincing evidence to third-party certification bodies is a huge challenge for China's forest carbon projects.

Common practice analysis is a necessary step in the analysis of additionality. Project participants need to analyze local forestry activities such as A/R, forest management and conservation; and provide transparent and verifiable evidence of how project activities differ from these universal activities. Otherwise, if the project activity is not significantly different from such general forestry activities, the project activity is not additional. In order to achieve the target of increasing 40 million hectares of forest area by 2020 and raising forest reserves by 1.3 billion cubic meters, the Chinese government and enterprises invest large amounts of money each year for A/R, forest management and conservation. Therefore, how to distinguish project activities from the general forestry activities invested by governments and enterprises and provide convincing evidence to third-party certification bodies is also a huge challenge for China's forest carbon projects.

6. Programs and methodologies are complex and inflexible

Both CDM A/R projects and forest carbon projects under the voluntary carbon standards require complex procedures, including project feasibility study demonstration, project development, independent third-party validation, application for registration, project implementation, monitoring, verification, and certification, so as to obtain tradable carbon sink. The CDM A/R project also needs to apply for approval letter to the NDRC through province, municipal, or autonomous region and the SFGA. Voluntary carbon projects also need to be filed for record keeping with the NDRC (2012).

Due to the particularity of forest itself, relevant rules and methodologies are much more complicated than the emission reduction projects in other sectors, making the development, implementation, and monitoring of forest carbon projects difficult, and the projects are subject to longer duration, higher transaction costs, lower carbon benefits, and higher risk. The carbon benefit is inconspicuous compared with investment. Moreover, due to the long period of forest carbon project, the carbon benefit will be gradually obtained over decades or even longer, making benefit lag far behind project investment. This greatly limits the enthusiasm of project participants in the CDM forest carbon project. The World Bank's BIOCF supported more than 20 CDM A/R projects. From project development to successful registration, it took 5.4 years on average before 2007, including 3.9 years of project development, 1.2 years of certification, 0.3 years of registration; the number is 2.9 years on average after 2007, including 1.4 years of project development, 1.1 years of certification, and 0.4 years of registration (The World Bank 2011). Previous projects have a longer development period because they do not have appropriate methodologies and related development experience. The "reforestation project at the Pearl River Basin of Guangxi in China" takes about 3 years from development to registration. The "Degraded Land Reforestation Project in Northwest Sichuan" takes about 4 years, and the "Degraded Land Reforestation Project in Northwest Guangxi" takes about 4 years.

For general international cooperation projects on forestry, there are also project design activities, but there are no strict requirements on land eligibility, project boundaries, baseline scenarios, additionality, greenhouse gas emissions, and leakage. In general, as long as relevant tasks are completed, the assessment can be cleared. However, the implementation of forest carbon project is different from the general international cooperation and domestic implementation projects. The requirements are strict and must be strictly implemented according to the project design. Otherwise, it may affect the baseline, additionality, and greenhouse gas emission sources, leakage, measurement, monitoring methods, and monitoring plans determined during project development. If this happens to the CDM A/R project, the PDD and monitoring plan must be revised and submitted to the CDM-EB for approval.

7. Rights of land use

The project participants have the ownership or use rights of the project land, which means that project participants have the right to carry out project activities on the project land. However, in many parts of China, the right to land use is unclear and affects the development and implementation of forest carbon projects.

Since the founding of new China, the collective forestry ownership system has undergone several changes, but there are still problems such as unclear property rights, failure to establish operating entities, inflexible operating mechanisms, and unreasonable distribution of interests. In June 2008, the State Council promulgated the "Opinions on Comprehensively Promoting the Reform of Collective Forest

Ownership System" to advance the reform of the collective forest ownership system in an all-round manner. While maintaining the ownership of collective forestland unchanged, the contractual forestland ownership shall be legally implemented to farmers of collective economic organizations through household contracts, thus establishing farmers contractual ownership to the use right of forestland. For forest carbon project implemented in China in the early stage, the collective forest ownership system reform was not available in the project area at the time of project development. The use right to forestland and the "four-side boundaries" were determined in the early 1980s through the "terms of reference for forestry." Most of the use rights are collectively owned by the villagers' group, and the project forestland use contract is signed by A/R entity and villager group. Due to the limitations of working methods and approaches at that time, the "four-side boundaries" of the forestland was not identified on the topographic map. There was only a simple text description, which is often inconsistent with the reality. The "four-side boundaries" was not clear enough. Before project implementation, the forestland was a long-term forest-free barren land. The potential benefits were yet to be tapped. Villagers did not care much about the ownership of land use rights. Therefore, there was seemingly no dispute. However, after the implementation of the project, villagers saw the added value and benefits of the land. There were disputes between villages. The project owners were not willing to bear the risks brought by A/R on the disputed forestland, which made the project impossible to implement. For example, the CDM A/R project "reforestation project at the Pearl River Basin of Guangxi in China" designed 147.5 ha of forestland, but the A/R can't be achieved due to such land disputes. The CDM A/R project "Degraded Land Reforestation Project in Northwest Guangxi" designed 1,052.8 ha of forestland, but was unable to implement A/R due to land disputes, and there were even ownership disputes between counties.

With the full implementation of the reform of China's collective forest rights system, the forestland that was originally used by the villagers' group was contracted to farmers. The part of forestland uses contract originally signed between the project owner and village group needs to be reestablished with forestland contractor. Some farmers took the opportunity to put forward more unreasonable requirements (such as requesting to increase the share or rent, requesting road repair for villages, etc.), resulting in the fact that project entity can't sign land use contracts with some farmers. For example, due to this reason, 173.3 ha of forestland designed for the "reforestation project at the Pearl River Basin of Guangxi in China" was not able to implement.

Since forest carbon projects often involve thousands of households, at the time of project review, some landowners or users of project land cannot fully determine whether they will participate in the forest carbon project. The CDM A/R project allows to provide proof of 2/3 area of ownership certificates during project review, and the remaining 1/3 is to be provided at the time of initial verification (CDM-ED 2008). However, forest carbon projects are generally not implemented on all plots at the same time, and a large number of plots need to be gradually implemented in the following years after the start of the project. The general situation in China is that farmers or village collectives are reluctant to sign cooperation agreements or land use rights transfer agreements in advance, they prefer signing such documents at the time of implementation. For forestland leased from village collective or farmer, if a lease agreement is signed and project won't start immediately, the project owner will have to bear the increase in project investment. For these reasons, it is difficult to meet the 2/3 area of ownership requirement for project review. At the time of the first verification, some project sites may not have some to-be-implemented project activities, which make it impossible to sign land use agreements.

8. Carbon pool

The carbon pool that can be measured and monitored by CDM A/R projects includes aboveground biomass, belowground biomass, dead wood, litter, and soil organic matter, but the carbon pool of wood product is not considered. Once trees are harvested, the carbon in the biomass is assumed to be immediately released into the atmosphere. In fact, after the forest is disturbed by the nature (such as forest fires, pests, rain, and snow, etc.) or human beings (such as harvesting), the removed biomass will be turned into wood products and continue to exist for a certain period of time (excluding logging and processed residues), depending on the type and use of wood products. In the case of sustainable operations, even when the carbon stock of forest biomass and other carbon pool no longer increase, the carbon stock of wood products produced through forest management is still increasing. For example, in bamboo forest, after 6–10 years of A/R, the biomass basically reaches a dynamic equilibrium status, but the carbon in the bamboo products produced by bamboo harvest every year can be preserved for a long time, that is, when the carbon in the biomass no longer increases, the carbon stock in bamboo product will continue to increase. If the carbon in the product is not considered, the project participants will lose a considerable amount of carbon sink.

9. Implementation of body

The China Clean Development Mechanism Project Management Measures stipulates that Chinese-funded and Chinese-holding enterprises in China can open up CDM projects to the world (NDRC 2011). Due to the following three reasons, it is still difficult for enterprises to take the lead in developing and implementing forest carbon projects. First, the land that meets the eligibility requirements of CDM A/R projects in China is generally located in old, small, remote, and poor areas, and the economy is underdeveloped. It is difficult to find capable enterprises to develop and implement CDM projects in the region. Second, with the exception of a small number of state-owned forestland, most of the land ownership that meets the eligibility requirements of the CDM A/R project is owned by the local community or individual. In particular, with the full implementation of the reform of collective forest rights system in recent years, more and more forestland use rights have been assigned to farmers, which has brought great difficulties for some nonlocal enterprises to participate in A/R in collective forest areas. Third, for a long time, China's A/R has been dominated by the government, especially in economically underdeveloped areas. Enterprises have invested less, and community cooperation in this regard is little. Community people doesn't have enough trust in enterprises, and they are worried that their due rights and interests can't be guaranteed.

1.2 Capacity Obstacles

The development of forest carbon projects requires extensive expertise and management capabilities, such as forestry, forestry survey planning and design, forestry policies and regulations, ecology, economics, law, carbon trading and related regulations, carbon measurement and monitoring methodology, financing, organization, and coordination. Therefore, it is often difficult to organize such a multidisciplinary team at the project location, and the cost of employing external experts is high. At present, the development of China's forest carbon project relies mainly on the technical support provided by a few international and domestic NGOs. The technical and capacity obstacles faced by forest carbon projects are mainly reflected in the following aspects:

Methodology development: Both the CDM and the voluntary forest carbon project with carbon criteria require using the approved methodologies. At present, the approved methodology for CDM A/R project basically met the needs of most A/R projects. However, there are few methodologies for forest management, vegetation restoration, and carbon projects reducing forest degradation that are suitable for China's national conditions and forest conditions. The development of methodology involves land qualification, project boundary, carbon pool selection, baseline scenario identification, additionality demonstration, carbon stock changes in various carbon pools, and measurement and monitoring of greenhouse gas emissions and leakage, risk assessment, etc. The development process is rather complicated, and there are very few domestic experts who can develop relevant methodologies.

The implementing body, local competent authorities, and technical personnel did not know enough about the rules and requirements of forest carbon project. Especially in areas where forest carbon projects have never been developed and implemented, people have a very superficial understanding of relevant rules, and the ideological concepts still remain in the conventional ideas and methods of A/R, thinking that forest carbon project is like "free treats" without making any contribution. Even after repeated training, some technical and managerial personnel still pay insufficient attention to relevant rules and requirements, which often leads to lots of repetitive work, so that project development and implementation often lag far behind the schedule, or causing increased monitoring tasks, which cause the project to lose its eligibility in severe cases. For example, regarding the "Small-scale Reforestation Landscape Restoration Project in Tengchong of Yunnan" registered by CCB, some A/R activities occurred outside the project boundary, which led to the revision of project design document during DOE verification; for the CDM A/R project "reforestation project at the Pearl River Basin of Guangxi in China," some plots didn't implement A/R according to the designed model, which has increased the monitoring workload. Most of the CDM and CCB A/R carbon projects implemented in China are affected by the erroneous inclusion of individual forestland into the project site, thus causing projects land to lose eligibility. At the time of the verification, they are required to revise project boundaries and PDDs.

The rules, guidelines, clarifications, methodologies, and tools of relevant carbon criteria are constantly being updated and such changes and updates need to be closely tracked. Domestic project owners and relevant professional and technical personnel seldom pay attention to these changes until being identified by independent third-party agencies during review or verification. They only begin to pay attention and adjust relevant contents, resulting in a great waste of manpower, financial resources, and time.

Compared with general forestry projects, one of the most prominent requirements of forest carbon project is measurable, reportable, and verifiable (MRV). To be "verifiable," any reasoning, judgment, and activity require transparent, verifiable evidence, not just a description. On one hand, there is no evidence in many cases. For example, many A/R methodologies are only applicable to degraded lands, and project participants must provide evidence that upcoming A/R projects are degraded or in the process of degradation, but many areas lack relevant research or degradation report on land classification. On the other, project owners and project participants lack the awareness of collecting and documenting relevant evidence. Some past activities, such as training, community interviews, project boundary identification, A/R activities (soil preparation, planting, fertilization, etc.), has no relevant evidence (such as photos, implementation contracts, reports, small-plot monitoring cards, etc.), causing difficulties to project validation and verification.

Shortage of reliable carbon measurement parameters. The forest carbon project must make an advance estimate of the project's expected carbon sink, and the important basis for the estimate is stand volume or biological growth equation. There are no such growth equations in many parts of China. Some literature only covers single-wood growth equations, when applied to stand level, there are no patterns regarding the number change of plants with respect to forest age increase. Although there are a large amount of sample plot data from forest resource surveys in various places, which serves as the basis for deducing relevant growth equation, but on one hand, it is difficult for project participants to obtain such data (confidential in China), on the other, there are very few sample plots for certain tree species stand, which can't meet the accuracy requirements. The equations concluded with such sample plot data represent the average condition in the region, and do not reflect the growth level of the environmental condition (such as degraded land) where the project site is located. For the latter, DOE often requires project participants to provide evidence that the growth equation used is appropriate for the degraded land in which a project is located.

For forest management carbon projects, the problems are much more serious. The estimation of low-yield forests in the baseline scenario, the growth estimation covers those after grazing exclusive, replanting, thinning, and rejuvenation, plus the mortality rate in baseline scenario and project scenario, there is almost no data to be referenced; it is difficult to obtain the type, use, and service life of wood products from thinning or harvesting. These have brought great challenges to the development of forest management carbon projects.

Insufficient experience in A/R. Several CDM A/R projects implemented in China have very poor site conditions due to severe land degradation. Due to the relatively small number of A/R tests and pilot studies on these difficult sites and lack of A/R experience, these projects have no basis for tree species selection and A/R technology, resulting in the unreasonable selection of A/R tree species and poor A/R site conditions. Some A/R projects failed or led to poor forest growth, causing a substantial increase of A/R costs and reduction in expected benefits. The enthusiasm and confidence of A/R entities in projects implementation are affected. For example, in the CDM A/R project "reforestation project at the Pearl River Basin of Guangxi in China" and "Degraded Land Reforestation Project in Northwest Guangxi," 469.8 and 367.7 ha of project land can't successfully implement A/R, due to the high altitude, steep slope, high wind and insufficient soil depth in mountainous area, accounting for 11.7% and 4.2% of the total project area, respectively. Even though some plots were reconditioned or replanted, the survival rate could not meet the requirements.

1.3 Investment and Capital Obstacles

(1) High investment, low benefit, and slow return

The A/R carbon project is the same as general A/R activities, with high cost and most of the investment is in the land, seedlings, forestland clearing, soil preparation, planting, and young forest rearing during A/R. The post-management cost is relatively smaller in the overall cost. But the benefits of timber as a major outcome can take years or decades or even hundreds of years to come. Ecological public welfare forests have less wood revenue and longer payback period due to harvesting restrictions. Therefore, in addition to short-rotation industrial raw material forests and fast-growing and high-yield timber forests, the return on investment for A/R is very low or even negative. At the same time, there are highly unpredictable risks of forest fires, pests, and diseases and natural disasters in A/R projects, plus low return on investment, thus banks are generally reluctant to grant project loans to A/R entities, except for a small number of industrial raw materials forests and fast-growing forests with high yields and short payback period. Provide. Under this circumstance, the project A/R funds often rely on the A/R entities themselves, which greatly affects the enthusiasm of A/R entities to develop A/R carbon projects.

(2) High transaction costs

The transaction cost of forest carbon projects involves project development costs (including PDD and methodology development when necessary), independent third-party certification fees, registration fees, monitoring fees, and independent third-party verification fees and issuance costs. The CDM A/R project supported by the World Bank's BIOCF has a project development cost of US$174,000 (small-scale project US$153,000), review fee US$16,500–45,000 (small-scale project

US$16,750–28,200), registration fee US$16,500–48,000 (no cost for small-scale projects), and verification fee US$14,300–53,200 (The World Bank 2011). From project development to registration, the transaction cost is US$203,000–493,000 (small-scale projects US$166,750–328,200). In addition, monitoring and verification cost take place every five years. The project development cost is the main part of transaction costs. The development cost per ton of carbon dioxide equivalent is US$0.40–3.70, and the average is about US$1.32, which is much higher than the US$0.78 for the wind power project, US$0.51 for biomass power generation and US$0.29 for hydropower. The development cost of A/R/ reforestation project accounts for about 0.5–20% of the total project cost, with an average of about 6%, accounting for about 30% of carbon income. Especially for small-scale A/R projects, the total development cost is not much lower than that of conventional projects, and the annual carbon is only 16,000 tons of carbon dioxide equivalent, and the average development cost per ton of carbon dioxide equivalent is US$1.50. Most project owners and participants do not have enough capacity to develop carbon projects and have to hire external experts to assist the development, resulting in a significant increase in development cost. As development capability is enhanced, development cost will decrease, and development cost for later projects is lower than that for earlier projects. For example, the development cost of the "Degraded Land Reforestation Project in the Northwestern Region of Guangxi" is about 30% lower than the "reforestation project at the Pearl River Basin of Guangxi in China,", which is the first project developed by the same team[9].

(3) Difficulty in using carbon income to solve investment barriers of A/R projects

Carbon income can increase the profitability of A/R projects to a certain extent, but it is difficult to solve the investment barriers faced by such projects in the following aspects: relatively high input costs, limited contribution of carbon income, and the carbon comes with tree growth after successful A/R and is subject to monitoring, verification, certification, and issuance; due to future international climate agreements, rules and carbon market uncertainties, the agreed period in the ERPA of A/R project is generally higher (mostly no more than 10 years); high transaction cost; the non-permanence makes market demand shrink, and carbon prices become low; most carbon buyers are reluctant to pay carbon funds in advance, so as to reduce the financial pressure on project participants. The carbon income of A/R projects are not even able to pay for project management cost and land opportunity cost (The World Bank 2011).

1.4 Market Obstacles

Due to the non-permanence of carbon in CDM A/R projects, carbon purchasers get non-permanent CERs, rather than permanent CERs generated by CDM projects in sectors such as industry and energy. To this end, the buyers must purchase additional

CERs to replace the invalid tCERs and lCERs after they lost their validities. This directly leads to economic and market consequences that are not conducive to A/R projects, namely low prices and the insufficient market for buyers. Given that such non-permanence, uncertainty, and leakage may bring greater risks to the EU ETS and the compliance of member states, as the largest purchaser of global emission reductions, EU ETS does not accept the carbon credit generated by forest carbon project, and does not allow European companies to use the carbon credit from forest carbon project to complete their committed emission reduction targets, thus further impacting the forest carbon market. Although EU member states are allowed to use carbon credits from certain forest carbon projects to help them meet the reduction committed under the Kyoto Protocol, few member states use it in practice.

The international community has recognized the impact of non-permanence on the forest carbon market and is also discussing alternatives during the negotiations for the second commitment period of the Kyoto Protocol, such as host country taking reversal responsibility for forest carbons projects, buffer carbon credits, carbon credits bank and insurance, etc. Some voluntary carbon standards, such as VCS and Panda standards, use buffer-based carbon credit approach to address such non-permanence. However, before the second commitment period is officially introduced, the market's confidence in forest carbon is hard to establish.

1.5 Risk Obstacles

(1) Fire risk

Affected by the monsoon climate, most of China is subject to clear dry and wet seasons, and forest fire risk is high in most areas during the dry season. For example, from 1999 to 2007, the average annual forest fire in China reached 340,000 ha, accounting for 0.12% of the forest area in the same period. In 2003, these figures top 1.12 million hectares, accounting for 0.66% of forest area. During this period, the average annual forest area affected in Heilongjiang Province accounted for 3.29% of the total, and the percentage is Inner Mongolia 0.96%, Guizhou 0.56%, Zhejiang 0.48%, Fujian 0.44%, and Guangxi was 0.40%. From 1999 to 2007, the annual forest loss from forest fire was about 2.04 million cubic meters, with the highest in 2006 topping 10 million cubic meters. The crediting period of forest carbon projects is usually more than 20 years, and the accrued proportion of damage period will be high. In the event of a forest fire, a significant portion of the accumulated carbon will be released into the atmosphere instantaneously and cause non-CO_2 greenhouse gas emissions. Therefore, forest fire is the biggest risk to China's forest carbon projects and the biggest challenge for forest carbon projects. Although there are no reports of forest fire in the forest carbon project implemented in China, the risks are always there.

(2) Extreme climate risk

China is a country with frequent extreme weather events. Rainfall, snow, and extreme droughts often cause significant loss to forest carbon projects. For example, in the CDM A/R project "Reforestation Project at the Pearl River Basin of Guangxi in China,", the rainfall, snow, and freezing disaster in early 2008 destroyed 595.1 ha of forest, and the 120.3 ha of forest, which were replanted after the disaster, was hit again in early 2011 and need to be replanted again; 197.8 ha of project stand suffered severe drought from 2009 to 2011 and had to be rebuilt. In the CDM A/R project "Degraded Land Reforestation Project in Northwest Guangxi,", 429.7 ha of project stand were destroyed during the snow and freezing disaster in early 2008; during 2009–2011, 645.8 ha of project forests suffered severe drought and has to be replanted times again. The CDM A/R project "Degraded Land Reforestation Project in Northwest Sichuan" was seriously affected by the "5.12" Wenchuan Earthquake in 2008 and its secondary disasters. The project sites completed in the spring of 2007–2008 were damaged to varying degrees. From 2009 to the spring of 2011, the area of replanting and rebuilding was 1,402.1 ha, accounting for 62.3% of the total. The large-scale of rebuilt forest stand due to these extreme weather events not only increased the cost of project participants but also affected the growth of forest stands and expected carbon sink, and it greatly affects the enthusiasm of project participants. Some farmers withdrew from the project as a result.

(3) Risk of land use change

After the project is implemented, if there are more profitable land use opportunities (such as mining, brick factory, planting cash crops, etc.), farmers often choose to withdraw from the project. For the CDM A/R project "Reforestation Project at the Pearl River Basin of Guangxi in China" and "Degraded Land Reforestation Project in Northwest Guangxi,", there are 37.5 ha and 116.8 ha of such withdrawn forestland, respectively, which makes it impossible to implement A/R projects.

2 Experience Sharing

2.1 Communication with National Authorities

China's "Operation and Management Methods for Clean Development Mechanism Project" stipulates that the qualifications, approval procedures, CER prices, and quantities of CDM projects must be supervised by the National CDM Project Review Board (NDRC 2011). At the same time, the CDM project must obtain the approval letter from the NDRC before submitting to the CDM-EB for registration. The approval letter must be reviewed by the SFGA and reported through local development and reform commissions. The approval process is a little complicated. In addition, for

CDM project with a start date of August 2, 2008 (including August 2), the NDRC must be notified within 6 months after the project start date. Therefore, during the preparation, the project participants must fully communicate with the NDRC and local development and reform commissions, the SFGA and the local forestry bureaus on the above issues, and reach consensus on project organization, income distribution mode, project management mode, and CER prices, so as to ensure that the project will pass the approval and begin implementation.

2.2 *Effective Management Structure, Giving Full Play to Local Forestry Authorities*

For a long time, local forestry authorities, especially county and township (town) forestry authorities have played a leading role in A/R. They are responsible for drafting A/R annual plan, organizing specific A/R, providing technical guidance and training, inspection and supervision, acceptance, and the issuance and management of forest ownership certificates. They have established long-term and close ties with local communities and farmers. They enjoy high prestige among the locals and are trusted by them. Moreover, the county-level forestry authorities manage local forestry archives, they are familiar with local environmental conditions, forest resources, and socioeconomic conditions, and have rich technologies and experience in A/R, forest management and forest protection, as well as cooperation with local communities. Therefore, making full use of the advantages of local forestry authorities will facilitate the smooth development of the project and its implementation, and may obtain government-matching funds, thus reducing related costs, and minimizing or eliminating related obstacles and risks. Several CDM and CCB A/R projects that are successfully implemented in China have established project management offices in provincial and project county forestry authorities to coordinate and organize project implementing entities, expert groups, farmer communities, and A/R entities, in an effort overcome various policy barriers, technical capacity barriers, investment barriers, and risk barriers encountered during development and implementation. All of these have enabled the project to be successfully developed and implemented. The commonly used project organization management structure is shown in Fig. 2.

(1) Responsibilities of the provincial project management office

Organize the application, implementation, and responsibility of the project with relevant state authorities.

Organize provincial-level training for county-level project offices, A/R entities/farmers, and coordinate the relationships between relevant participants involved in project implementation activities.

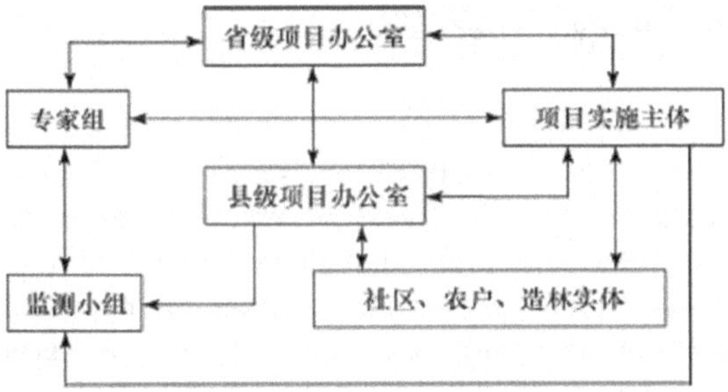

专家组	Expert group
监测小组	Monitoring group
省级项目办公室	Provincial project office
县级项目办公室	County project office
项目实施主体	Project implementing body
社区、农户、造林实体	Community, farmers, A/R entity

Fig. 2 Organization and management structure of forest carbon project

Review annual operation design of the project, check the progress and quality of A/R; coordinate, and solve major technical and economic problems in the project implementation process.

Review annual work report completed by the project implementing entity.

Review the annual monitoring plan, coordinate the expert group to complete, and review project-monitoring report.

Ensure the quality of monitoring.

(2) Responsibilities of the county-level project management office

Formulate an annual implementation plan, conduct annual job design, lead, and supervise the implementation of the project.

Coordinate the relationship between the forest farms/companies, farmers, and related departments involved in the project; solve the technical and economic problems in the project implementation process.

Conduct publicity activities on carbon projects.

Conduct technical training for forest farms/companies and farmers participating in the project, provide technical guidance, inspection, and supervision for various A/R activities, and responsible for the preparation and reporting of various materials required for the implementation process.

Organize technical personnel to cooperate with the expert group and project monitoring team to carry out relevant investigation and monitoring work.

Establish and manage project files at the county level (including farmer groups), and guide the company/forest farm to manage the file properly.

Manage and supervise carbon revenues to ensure that distribution is reasonable and in place.

2.3 Multidisciplinary Expert Team

In China, a forest carbon project is a new type of project that has emerged in recent years. The rules, procedures, and methodologies are complex and the review process is strict. It involves not only A/R, forest protection, forest survey, planning and design, and forestry-related disciplines, but also international rules and procedures, methodologies, sociology, management, law, carbon trading, and domestic policies. Therefore, it is necessary to organize a comprehensive expert group with expertise in various subject areas to participate in project development and provide technical guidance for project implementation and monitoring, so as to ensure successful project development and implementation.

For example, the "Reforestation project at the Pearl River Basin of Guangxi in China" has, from the outset, established an expert group consisting of international, national, and local experts from all disciplines, who understands international rules and domestic policies and are familiar with project background, with the World Bank providing legal and carbon trading support. Local experts are responsible for A/R design, and under the guidance of international and domestic experts, carry out project site selection, qualification demonstration, project boundary determination, baseline survey, community survey, investment analysis, participatory rural appraisal, and environmental impact analysis; major technical issues, such as methodological development, additionality arguments, leakage, carbon measurement, and monitoring programs are dealt with by international and national experts with the cooperation and support of local experts.

In this way, the expertise of international, national, and local experts have been fully utilized to complement each other and developed the world's first CDM A/R methodology and project. During the project implementation and monitoring process, local experts are responsible for formulating project inspection and acceptance methods, project implementation guidelines, project monitoring manuals, project training materials, forest growth standards; they also provide technical training for the project, and assist provincial project offices to solve major technical problems in project implementation, conduct quality inspection of A/R survival rate and preservation rate of county-level project offices and A/R entities, review and verify the reliability and accuracy of the monitoring team, field data and laboratory data of verification workgroup, and analyze and study the monitoring results and draft monitoring reports. International and national experts provide technical advice for project monitoring, especially monitoring changes in carbon reserve in specified plots.

2.4 Sound Cooperation Mode

Among the forest and forested land in China, 60% are collectively owned; of the collectively-owned forested land, 47% are collectively operated and 53% are operated by an individual (SFGA 2009). However, it can be foreseen that with the full implementation of the reform of collective forest ownership right, the proportion of individual operation will also increase. Therefore, establishing a cooperation model featuring sound input and benefit distribution with land users not only meets the requirements of China's climate change authorities for the qualifications of project participants, but also a policy obstacle that should be solved before developing forest carbon projects. It is also the key to project sustainability. At present, there are three modes of cooperation in China's forest carbon projects:

(1) Land Lease

Project owners and other entities lease land from local farmers, communities, or other landowners for forest carbon projects. The landowners receive rental income, project owners, and other entities pay for land rents, independent investments, and project development, and are entitled to all forest products and carbon revenue during the lease period. Under this model, farmers' income is low, and participation is not conducive to the sustainable management and protection of forests, and the risks are high.

(2) Shareholding cooperation

The project owner and other entities cooperate with landowners in a shareholding arrangement. The local landowners provide the land; the project owner and other entities invest in and carry out project activities and bear the risks of nature and investment, and the project owner is responsible for the development, registration, and monitoring of the project. Project owners and other entities enter into contracts with landowners to clarify management responsibilities, inputs, and income distribution. Revenues include wood, non-timber forest products, and carbon sink, and different incomes can apply different distribution ratios. In addition, local farmers can also earn labor income by participating in project activities. This model fully reflects the extensive participation of the community. The community benefits are directly related to the quality of project implementation, which enhances the enthusiasm of the community to participate and manage the project well, thus it is considered to be one of the effective models. Valid contracts by some projects are tripartite agreements involving entities, landowners, and local county-level forestry authorities.

(3) Independent participation of farmers or communities

Individual farmers or communities invest in project activities on their own land. Local forestry authorities provide relevant technical training and services to guide project implementation and quality control. Forest products and carbon revenues all belong to farmers or communities. Farmers and community authorized project owners to carry out project development, registration, monitoring, and carbon negotiation, sign purchase agreement and engage in sales. This model has the highest level of community participation, but the enthusiasm of project owners is limited. Therefore, this model is only applicable to a small number of activities in a project.

In the above model, a project often involves multiple local entities (forest farms, forest companies, etc.). In this case, these entities can work separately with local landowners on the land lease or shareholding cooperation. The project entities may elect one entity with economic and technical strength as project owner or implementation entity. It is authorized by other entities to apply for an approval letter from the NDRC, engage in project development, registration, monitoring and carbon negotiation, and sign purchase agreement and sales. This is in line with the requirements of the "Project Management and Operation Method for Clean Development Mechanism,", and it is easy to operate and is easy to be accepted by entities. This method is considered to be an effective way and should be promoted in China.

2.5 Systematic Training and Publicity, Improving Management and Technology Capacity

In China, forest carbon is new to most management and technical personnel, especially in places where forest carbon projects have never been carried out, as well as for such personnel at the grass-root level. Most of them only get some information from media reports. Little is known about the connotation and mechanism of forest carbon, the rules, methods, and procedures for applying and implementing forest carbon projects and related methodologies. Farmers and communities have little knowledge of forest carbon. Therefore, different types of publicity and training for different audiences are one of the important means to break down obstacles in project development and implementation.

Provide training for provincial and county forestry authorities and project entity management personnel, local scientific research institutes, university science and technology personnel, and local forestry technicians. The training contents include climate change and its impact, emission reduction implication and negotiation progress, origin, role and connotation of forest carbons, international rules, methods, procedures, and domestic policies for forest carbon projects under CDM and voluntary carbon standards, forest carbon project development processes and requirements, carbon trading rules, and domestic and international forest carbon markets, etc.

Conduct technical training for local scientific research institutes, university science and technology personnel and local forestry technicians, especially those involved in project development and implementation. The training content includes basic requirements and preparation of project documents, related methodologies and methodological tools, project boundary determination, land eligibility demonstration and identification methods, baseline survey methods, community survey methods (participatory rural appraisal), A/R, forest management and management design requirements, socioeconomic and environmental impact analysis, additionality argumentation, measurement and monitoring methods for changes in carbon reserve, identification, and measurement and monitoring of greenhouse gas emissions and leakage, etc. Such training covers classroom training, field training, and internship.

In conjunction with baseline surveys and participatory rural appraisals, publicity and training will be provided to communities to develop a clear understanding of the role of forest carbon, forest carbon trading requirements and risk, community participation methods, responsibilities, and interests, so as to raise community awareness to the project.

Organize the management and technical personnel involved in project development design and implementation to participate in international training, communicate more with international experts, and learn from the experiences and lessons learned of other countries in the development and implementation of forest carbon projects.

2.6 *Participatory Design and Monitoring, Enhancing Communication, and Coordination with Communities*

Participatory rural appraisal is a method for analyzing rural problems and working with local stakeholders for a solution. It uses a wide range of visualization methods for population analysis, so as to address existing social and environmental issues from space and time perspectives.

Participatory rural appraisal plays an important role in the development of forest carbon projects. First of all, forest carbon projects often involve rural communities in a wide range of areas. The natural and socioeconomic conditions and the socioeconomic and environmental problems are complex and diverse. It is necessary to adopt participatory decision-making and planning mechanisms to give stakeholders the right to know and participate, as well as decision-making and supervisory powers, so as to enable them to participate in the decision-making, design, implementation, and supervision of project activities and obtain benefit, especially the participation of vulnerable groups including women and the poor. This is also why CDM A/R project activities and procedures require soliciting and considering stakeholder input on project activities. Participatory rural appraisal is an effective way to fully collect stakeholder opinion. Second, a participatory rural appraisal is also an effective way to understand the credible land use patterns and forest management activities of local communities and stakeholder groups in the absence of proposed forest carbon activities. Therefore, the information gathered through participatory rural appraisals can play an important role in determining alternative land use or forest management options for project activities, demonstrating baseline scenarios, and additionality of project activities. Third, through participatory rural appraisals, the socioeconomic status of the communities involved in forest carbon activities can be understood and assessed, and the socioeconomic and environmental impacts of the proposed project activities can be analyzed, which is required by relevant rules of forest carbon projects.

Therefore, for forest carbon projects, the purpose of participatory rural appraisals is as follows: obtain basic documents, data and information on the natural and socioeconomic conditions of the proposed project area and its surrounding areas (such as climate, soil, vegetation, and ecosystems, biodiversity, land use status, forest management activities, socioeconomic activities, history and current status of socioeconomic conditions), understand the main socioeconomic and environmental issues, collect stakeholder willingness in land use and forest management and demand for participating in projects, analyze the potential socioeconomic and environmental impact of the project activities, thereby providing a basis for baseline scenario identification, additionality argumentation, socioeconomic and environmental impact analysis, and project participants' opinions, which are required for the development of forest carbon projects.

After the project design is completed, each community will be presented with the types of project activities, the concerned land, technical measures, and implementation time, through various media, such as text introduction, bulletin board, publicity poster, broadcast, etc. The opinions and suggestions of the community concerned will be solicited.

During the implementation of the project, unimpeded communication procedures and channels need to be established to ensure that local communities and other project participants can reflect their difficulties, complaints, opinions, and suggestions during project development and implementation. Channels to communicate with the community include:

Collect complaints, opinions, and suggestions when local community participates in project work;

The forest guards keep regular communication and exchange with community farmers during forest protection process;

Local (county and township level) forestry authorities regularly visit the community involved in the project to solicit relevant opinions and suggestions.

Written replies will be given within 30 days after collecting complaints, opinions, and suggestions. In order to gain an in-depth understanding of the impact of forest carbon projects on local communities and farmers, track their socioeconomic changes, understand the difficulties and demands of local communities and farmers in the implementation of the project, and solicit opinions and suggestions, participatory rural appraisal method will be used to monitor the impact on communities while monitoring carbon stock change every time.

2.7 Strict Quality Assurance and Quality Control

Quality control is a series of routine technical procedures to detect and control project activities and to investigate, measure and monitor quality, including checking and ensuring that project activities are implemented according to project design, as well as true, correct and complete data of the survey, measurement, and monitoring. It also ensures that errors and omissions are discovered and coped with, all quality control activities are recorded and archived. Quality assurance is a planned evaluation system. It is an independent third-party review on the basis of project activities, monitoring and quality control by project participants. By verifying the quality of data, project activities are ensured, and the results of survey, measurement, and monitoring are accurate and reliable. Project participants should develop relevant quality assurance and quality control procedures that include, at a minimum, data collection, survey and measurement, data validation, data entry and analysis, and data archiving. Quality assurance and quality control include:

Project development: methodology selection, methodology, and application of methodological tools, baseline survey, community survey, land eligibility, project boundary determination, project activity design, project baseline carbon, and project carbon sink;

Project implementation: type of project activity implemented, technology employed, geographic boundaries, and quality of implementation;

Monitoring of project carbon sink: measurement and monitoring of project boundary and carbon stock changes.

2.8 Timely Collection of Evidence and Archive Management

Through training, project participants and management and technical staff are informed of the various evidence that should be gathered and archived from time to time, such as:

The basic information (such as geographic location, area, land ownership, etc.) of each subplot is recorded by a monitoring card designed for each. All the A/R and forest management activities and their dates that occur in each plot are filled in by a designated person and signed and archived by the person filling the form (see the appendix of this chapter for a sample form). At the time of verification, the plot-monitoring card can serve as evidence for A/R and forest management activities.

Develop and document standard operating procedures for baseline survey, community survey, fixed sample site survey and monitoring and various other surveys and measurements; record and archive original records and data, with the date of survey and signature of surveyors.

Training notification, check-in forms, training materials, pictures of classroom training and field training internships, training performance records, and training certificates are issued.

Meeting notification, check-in forms, and photos of various project-related formal and informal meetings.

Field photos (close-up and panorama) of each project plot, live photos of various surveys and measurements.

Extensively collect and archive a variety of literature, drawings, and data, as well as national policy documents, laws and regulations and technical standards, various data sources, and copies referenced in project implementation documents and monitoring reports.

Publicity materials related to the project.

Various agreements, contracts, including technical services and expert contracts.

The above evidence must be filed in both electronic and printout formats and backed up in different locations (such as provincial and county project management offices, project participants, etc.). All electronic data and reports must be backed up through permanent storage carriers such as CD-ROMs, and the backup copies of these discs will be stored in different locations. For documents and electronic data, one personnel shall be assigned to take care of it.

3 Recommendations

3.1 Policy Recommendation

(1) Recommendation on international policy

In the second commitment period of the Kyoto Protocol or in future international climate agreements, the rules for forest carbon projects should be revised accordingly, including:

Improve forest management, conservation and increase forest carbon stock, prevent forest degradation and deforestation, and restoration of vegetation shall be included in eligible carbon trading of forest carbon projects.

Open up land eligibility conditions for CDM A/R projects. For example, the land is a no-forest area at least 10 years before A/R, which actually requires a new definition of reforestation.

Abandon the non-permanence solution adopted by CDM A/R project of the Kyoto Protocol's first commitment period, using methods to raise forest carbon prices and tradability and expand forest carbon market, such as host country undertaking forest carbon reversal, buffering carbon credits, carbon credits warehouse or insurance.

Give the project boundaries more flexibility. For example, when determining project boundaries, in areas with complex terrain or severely fragmented land, project participants are allowed to include a small number of unqualified plots within the project boundary, and the actual project area is calculated using a conservative discounting method. In the project validation phase, the requirement for the ownership certificate is reduced for 2/3 of the land. Only the ownership certificate at project commencement is required during verification, and for land that hasn't carried out activities within the project boundary, no proof of ownership is required at the time.

Simplify project rules, methods, and procedures; adopt more default methods to further simplify methodologies; reduce eligibility requirements for small-scale A/R projects.

The carbon pool of wood product is included in the measurement of forest carbon.

(2) Recommendation on domestic policy

As the main body of terrestrial ecosystems, forests provide a variety of products (wood, fiber and non-wood forest products) and services for people's production and life (maintaining water and soil, conserving water sources, reducing flood peaks, preventing wind and sand, regulating climate, conserving biodiversity, providing recreational sites, etc.). It is a critical guarantee for ecological security. Therefore, forest carbon projects can not only contribute to mitigating climate change but also reduce the vulnerability to climate change and enhance adaptability by increasing the quantity and quality of forests. At the same time, it can improve the environment and raise community income. To this end, the following recommendations are made for relevant domestic policies:

The forest carbon project shall be made a priority for international and domestic carbon trading.

In the development of domestic carbon trading rules, the experience of CDM A/R projects shall be learned from to improve on deficiencies (such as the aforementioned recommendations for international rules).

Forest carbon methodologies and methodological tools in relation to international and domestic voluntary carbon standards shall be recognized at the national level (such as VCS, Carbon Fix Standard, panda standards, etc.). Simplify and revise as necessary.

Encourage and provide a necessary fund to support the development of forest carbon methodology suitable for China, especially regarding vegetation restoration and wetland restoration.

Simplify additionality requirements and formulate additional criteria for judging China's national conditions, social and forest status. For example, for the construction and management of ecological forests for public welfare with low benefit, and for A/R or vegetation restoration projects on extremely difficult sites (severe desertification or desertification land, rocky desertification land, severe saline-alkali land, areas with annual precipitation below 600 mm, mining area reclamation), the source of investment can be required that government subsidies or public capital inputs shall not exceed a certain proportion (such as 30%) of total investment during the construction period; for short-rotation industrial raw material forests or fast-growing high-yield timber forests, or forest carbon projects with a rotation period of $\leq$20 years, the internal return on investment shall not exceed certain level (such as 8%).

Encourage the adoption of multiple benefit criteria (such as CCB) to enhance the social and environmental benefits of the project.

Further, promote the reform of collectively owned forest right system and reduce the risk of land ownership disputes.

Allow local forestry authorities to directly apply for and implement international forest carbon projects.

3.2 Technical Recommendation

More efforts shall be given to publicity and technical training of forest carbon projects, so as to enhance the understanding of the concept, origin, and rules and procedures of forest carbon on the part of forestry management authorities and company management personnel. Cultivate an expert team in each province, city, and autonomous region that can understand and track international and domestic carbon trading rules, and independently develop and monitor forest carbon projects.

Project shall be carried out in studying forest growth equations, biomass equations, loss rate, and other carbon measurement-related parameters under different types of forests, different site conditions, and stand conditions.

Projects shall be carried out to further study the technical methods and Low-function forest transformation techniques for A/R on difficult sites.

China has conducted seven consecutive forest resource inventory surveys and multiple forest management surveys. The forestry authorities at all levels have abundant survey data and maps. Many of the data are very useful for the development and monitoring of forest carbon projects, but these data are often difficult to access by project participants or obtained at a higher cost. Therefore, it is necessary to establish a sharing mechanism for forest data.

References

CDM-EB (2008) Guidance on application of the definition of the project boundary to A/R CDM project activity (Version 01). In: Report of the forty-fourth meeting of the executive board: annex 16, 26–28 Nov 2008

NDRC (2012) Provisional management method for trade in voluntary GHG emission reduction. NDRC documents (FGQH [2012] No. 1668), 13 June 2012. www.dtyjix.ong/zhengcefagui-157-1-2.shtml

NDRC, MOST, Ministry of Foreign Affairs, Ministry of Finance (2011) Operation and management methods for clean development mechanism project (revised), 22 Sept 2011. http://www.gov.cn/flfg/2011-09/22/content_1954044.htm

SFGA (2009) China forestry resource report—7th national forestry resource inventory. China Forestry Press, Beijing

The World Bank (2011) BioCarbon fund experience: insights from A/R and reforestation clean development mechanism projects

UNFCCC (2005a) Decision 5/CMP.1. modalities and procedures for A/R and reforestation project activities under the clean development mechanism in the first commitment period of the Kyoto Protocol: part two: action taken by the conference of the parties serving as the meeting of the parties to the Kyoto Protocol at its first session. FCCC/KP/CMP/2005/8/Add.1

UNFCCC (2005b) Decision 16/CMP.1. Land use, land use change and forestry. In: report of the conference of the parties serving as the meeting of the parties to the Kyoto Protocol on its first session. Addendum: part two: action taken by the conference of the parties serving as the meeting of the parties to the Kyoto Protocol at its first session. FCCC/KP/CMP/2005/8/Add.3

Zhang X, Shuhong W (2010) Theory and practice of forestry carbon project. China Forestry Press, Beijing

Zhang Z, Zhang X, Zhu J et al (2009) Study on the carbon cost of CDM A/R and reforestation—taking the CDM reforestation project at Guangxi Pearl Delta River Basin as an example 5(6):348–356

Appendix

Table 1 Sample form of A/R monitoring card

Monitoring item	Content of record			Notes
Geographical location	Plot number			
	County			
	Town (township)			
	Village			
	compartment number			
	Sub- compartment number			
Land ownership				
Type of bamboo	Bamboo species 1			
	Bamboo species 2			
	Type of forest mix and proportion			
Area/hectare	Design area			
	Actual operation area			
Boundary monitoring		Monitoring date	Changes as compared to previous monitoring	Project boundary coordinates attached
	1st time			
	2nd time			
	3rd time			
	4th time			
	5th time			
	6th time			
Forest land clearing	Date			
	Approach			
	Specification			
Land preparation	Date			Construction contract kept
	Approach			
	Specification			
Planting	Date			Construction contract kept
	Number of trees/hectare			
Fertilization	Date	Type	Fertilizer quantity	Photos and purchasing invoice kept
Survival rate and	Date	Survival rate/retention rate		

Z. Lu et al. (eds.), *Forest Carbon Practices and Low Carbon Development in China*, https://doi.org/10.1007/978-981-13-7364-0

retain rate inspection				Provide survey approach	
Replanting	Date	Bamboo type	Number of trees/hectare	Construction contract kept	
Tending and management	Date	Content	Approach and specification		
Case of pest and disease	Time				Identify the affected boundary and GPS coordinates
	Name of pest/disease				
	Affected area/hectare				
	Damage				
	Prevention approach				
	Prevention result				
Case of fire	Time				Identify the affected boundary and GPS coordinates
	Affected area/hectare				
	Type of fire				
	damage				
Harvest	Date				
	Approach				
	Quantity of harvest				
	Purpose				
Other management event and incident	Date	Activity and event description			

FSC
www.fsc.org
MIX
Papier | Fördert
gute Waldnutzung
FSC® C083411

Kolibri 360 GmbH
Ferdinand-Jühlke-Straße 7
99095 Erfurt, Deutschland
produktsicherheit@kolibri360.de